JONES AND BARTLETT SERIES IN BIOMED
SERIES EDITOR JULES J. BERMAN

Perl Programming

for Medicine and Biology

Jules J. Berman

JONES AND BARTLETT PUBLISHERS
Sudbury, Massachusetts
BOSTON TORONTO LONDON SINGAPORE

World Headquarters
Jones and Bartlett Publishers
40 Tall Pine Drive
Sudbury, MA 01776
978-443-5000
info@jbpub.com
www.jbpub.com

Jones and Bartlett Publishers
Canada
6339 Ormindale Way
Mississauga, Ontario L5V 1J2
CANADA

Jones and Bartlett Publishers
International
Barb House, Barb Mews
London W6 7PA
UK

Jones and Bartlett's books and products are available through most bookstores and online booksellers. To contact Jones and Bartlett Publishers directly, call 800-832-0034, fax 978-443-8000, or visit our website, www.jbpub.com.

Substantial discounts on bulk quantities of Jones and Bartlett's publications are available to corporations, professional associations, and other qualified organizations. For details and specific discount information, contact the special sales department at Jones and Bartlett via the above contact information or send an email to specialsales@jbpub.com.

Production Credits
Acquisitions Editor, Science: Cathleen Sether
Acquisitions Editor, Science: Shoshanna Grossman
Managing Editor, Science: Dean W. DeChambeau
Editorial Assistant: Molly Steinbach
Production Editor: Tracey Chapman
Senior Marketing Manager: Andrea DeFronzo
Manufacturing Buyer: Therese Connell
Composition: Arlene Apone
Cover Design: Anne Spencer
Cover Image: © Galina Barskaya/ShutterStock, Inc.
Printing and Binding: Malloy, Inc.
Cover Printing: Malloy, Inc.

Library of Congress Cataloging-in-Publication Data
Berman, Jules J.
Perl programming for medicine and biology / Jules J Berman. -- 1st ed.
p. cm.
Includes bibliographical references and index.
ISBN-13: 978-0-7637-4333-8 (pbk.)
ISBN-10: 0-7637-4333-X (pbk.)
1. Perl (Computer program language) 2. Medicine--Data processing. 3. Biology--Data processing. I. Title.
R858.B472 2007
610.285--dc22

2006037970

6048

Printed in the United States of America
11 10 09 08 07 10 9 8 7 6 5 4 3 2 1

Contents

Preface

0.1. BACKGROUND

Readers who are 20-something might find this hard to believe, but just a few decades ago, almost no biomedical professionals touch-typed much of anything. Manuscripts were handwritten (the Latin root of "manuscript" means "written by hand") and passed to secretaries for typing. Very few of us kept an electronic typewriter in our offices, and it was a rarity to see office computers until the late 1980s. For most of the 1970s, electronic typewriters had no memory. Each successive draft of a manuscript was laboriously retyped.

Nobody in the 1980s imagined that professionals throughout the world would choose to type ceaselessly, day and night, at home and in the office, on all manner of input devices. Collectively, we e-mail, blog, and text-message (now a verb) billions and billions of words. It is hard to find a biomedical professional who is not a prodigious typist.

In the past, very few professionals programmed. If they needed a software application, they would shop for a software application. If they found no appropriate software application on the market, they would seek the services of a professional programmer to write a software application.

In the mid-1990s, several high-level programming languages became freely available to the public. These languages were less fussy than earlier languages. Soon programmers were writing software in a fraction of the time and with much less effort than in the past. In addition to being easier to use, these newer languages were easier to learn.

Biomedical professionals are beginning to write their own programs. Many professionals believe that you cannot master collections of biomedical data if you cannot write your own programs. Obtaining basic competency in scripting languages (such as Perl, Python, or Ruby) can be easier than learning a new word-processing, spreadsheet, or statistics application. Increasingly, healthcare workers are finding that programming is an enjoyable and liberating experience.

The assumption underlying this book is that every professional will soon be programming. Most of the common computational tasks in a hospital and medical research setting can be done with short scripts of 10–20 lines, composed hastily, in a few minutes (much like e-mail).

The most common biomedical computational tasks involve collecting data, organizing data, searching data, or converting data to a desired format. All of these tasks, and many more, are easy to accomplish with Perl.

0.2. BIOMEDICAL COMPUTER PROGRAMMING

The goal of learning how to program is to develop computational self-reliance within your particular realm of interest. You will soon be programming your own short utilities (List 0.2.1). Large, complex applications will remain in the jurisdiction of full-time programmers and software-development teams. As a new programmer, you will benefit from an

LIST 0.2.1. SOME BIOMEDICAL TASKS THAT CAN BE ACCOMPLISHED WITH PERL

- Statistics
- Mathematical computations
- Mathematical modeling
- Web protocols (for example, http and ftp)
- Cryptographic techniques (for example, one-way hashes and encryptions/decryptions)
- Integrating data (for example, combining data from heterogeneous databases)
- Glue functions (for example, calling subroutines written in C)
- Digital Signal Processing (including image analysis)
- Bioinformatics methods (for example, calling Blast and FASTA)
- Database interfaces
- Remote procedure calls and distributed computing
- Software agents (via Web services, GRID, SOAP [see Glossary], and related protocols)
- Transformations to and from XML
- XML data queries
- Logical annotation of data (for example, RDF)

expanding awareness of dozens of dependable, free, open source applications and utilities. Non-programmers typically lack the self-confidence to try command-line utilities. Once you begin to program, you will never again fear a command line. Programmers, even part-time programmers, have many options unavailable to non-programmers.

By now, you have probably decided that you would like to learn to program. But are you qualified to learn a programming language? Do you have what it takes? The three essential qualities of a good programmer are: laziness, impatience, and hubris. If you're constantly trying to find an easier, faster way of working, and if you believe that the world owes you fame and fortune, then programming is just the ticket.

0.3. CHOOSING A PROGRAMMING LANGUAGE

What programming language must you learn? There are actually quite a few excellent high-level programming languages that are available at no cost. It makes little difference which language you choose. I have never encountered anyone who regretted his or her choice of programming language. However, there is a set of features that make some languages particularly suitable for biomedical professionals (List 0.3.1).

LIST 0.3.1. DESIRABLE FEATURES IN A PROGRAMMING LANGUAGE

- Short learning curve. Impatience is a virtue.
- Simple command syntax that resembles human language as much as is feasible.
- No-cost language that is widely available and that operates on any computer operating system.
- Friendly user community that helps newbies and that includes biomedical professionals like you.
- Cornucopia of instructional resources (books and Web pages) that explain every aspect of the language.
- Large library of freely available modules that can be called from your programs to perform highly specialized computational tasks.
- A forgiving programming environment that never crashes your computer and that includes useful compile-time error messages, often advising what needs to be changed to correct a bug.
- Built-in regular expressions. Regular expressions are a syntax for describing character patterns in text. A language that does not have regular expressions will have very little value for biomedical professionals.

Does any programming language have all these features? In my opinion, Perl, Python, and Ruby fit the bill. To some extent, this short list is a matter of personal taste and some readers would add Basic, Java, C, PHP, Smalltalk, Eiffel, Ada, MUMPS (now known as M), and others. Only Perl, Python, Ruby, and Java are supported by large, active biomedical user communities. Java, in my opinion, is a difficult language to learn and to use. Successful Java programmers tend to be full-time computer professionals. If you are primarily a healthcare professional, Java is probably not your best choice for a programming language. Among Perl, Python, and Ruby, Perl has the largest number of users in the biomedical community and has far-and-away the largest number of available resources (List 0.3.2).

Some people associate Perl with UNIX/Linux operating systems. Actually, Perl works just fine on Windows®. All of the Perl scripts in this book were tested in Windows®. It is impossible to know which programming language will be most popular ten years from this writing, but it is safe to say that the ghost of Perl will haunt the syntax and command logic for future

LIST 0.3.2. REASONS TO PROGRAM IN PERL

- Perl can be obtained at no cost.
- Perl is available for virtually every operating system and comes bundled into UNIX and Linux distributions.
- Perl is the most popular language in bioinformatics.
- It takes just a few hours to learn enough Perl to write your own biomedical programs.
- Perl programs tend to be much shorter and easier to understand than programs written in C or Java.
- A Perl script written for your computer will probably work on any other computer loaded with a Perl interpreter, even if that computer has a different operating system.
- Unlike C and C++, Perl comes with native pattern-matching commands (so-called regular expressions) that are used in virtually every program in the field of biomedicine.
- There are many thousands of Perl modules that are freely available from the Comprehensive Perl Archive Network (*www.cpan.org*). These modules perform a wide range of useful operations and can extend the functionality of your own programs.
- Perl code can be written in a manner that looks much like simple narrative text (if you make the effort), making it easy for others to read.
- Once you have learned Perl, you can migrate to almost any other programming language with ease.

generations of programming languages. For now, if you know Perl, you can easily understand short programs written in Python and Ruby.

This book is written for students, biologists, and medical professionals who want to program, and for programmers who want to learn something about biomedicine.

0.4. *BIOMEDICAL INFORMATICS*, AN OPTIONAL COMPANION BOOK FOCUSED ON BIOMEDICAL DATA

Jones and Bartlett published my first single-author book, *Biomedical Informatics* (2007). This book covers the entire field of biomedical informatics and stresses the regulations, ethics, and procedures for collecting, organizing, and sharing biomedical data. Readers of *Biomedical Informatics* were encouraged to learn a programming language, and an introduction to Perl programming was provided. Chapter 1 of *Perl Programminng for Biology and Medicine* is virtually the same as Chapter 5 of *Biomedical Informatics*. Some of the Perl scripts included in *Biomedical Informatics* are included here.

The purpose of *Biomedical Informatics* was to provide an in-depth understanding of the problems that face life science researchers and healthcare professionals. The purpose of *Perl Programming for Biology and Medicine* is to emphasize solutions. Readers of this book may seek detailed discussions of biomedical data resources, data standards, data organization, medico-legal and ethical conduct for data miners, and grants-related data-sharing responsibilities. If so, *Biomedical Informatics* is a useful companion text.

0.5. HOW TO READ THIS BOOK

Each chapter of *Perl Programming for Biology and Medicine* begins with a non-technical "Background" section that explains the biomedical issues underlying the Perl scripts that follow. The introductory remarks in "Background" are intended to inspire the readers with sufficient curiosity to read through the chapter. The last sentence of each "Background" section specifies the purpose of the chapter. Readers might want to approach this book by first reading the "Background" sections from every chapter, and, for in-depth study, selecting just those chapters that address their current professional goals.

The first two chapters may be difficult for readers with no background in programming. Readers might prefer to read these chapters along with an introductory Perl book, such as *Learning Perl* by Randal L. Schwartz and Tom Christiansen, or *Perl Core Language* by Steven Holzner.

The remaining chapters are organized by increasing complexity and subtlety of biomedical issues. In general, a biomedical scientist will find the first chapters difficult and the later chapters relatively easy. A programmer, who might know nothing of the issues that intrigue biomedical professionals, will find that the chapters start off very easy, but become quite difficult by the book's end. In either case, I am certain that every reader will encounter sections that are too difficult to understand on a first reading. Readers should feel free to skip difficult sections. The book is written with an extensive glossary that should help readers understand the text even when some sections are skipped.

All the Perl scripts included in the book implement basic algorithms. You can modify these basic scripts to suit your particular purposes. All scripts shown in the book can be downloaded from the Jones and Bartlett web site to copy and paste into your own programs. The Jones and Bartlett site will also provide bug notices and corrections, as needed.

The References section is an annotated list of every citation in the book. Each citation includes a comment explaining the relevance of the work to biomedicine. Many of the references link to useful Web resources (documents or software). These references provide a guide to some of the most important literature in biomedical informatics.

Resources list many free, open source software utilities and biomedical data resources, and describes how they are obtained and how they are used. The Resource contents are restricted to products that are mature, tested, and implemented by large user communities.

The glossary is written in an informal style intended to sustain your interest, and it can be read as a stand-alone text. It is an extensive list of terms related to computational biomedicine. Most of the Perl commands found in the text are explained in the glossary. In many cases, a Perl snippet demonstrates the command. The field of biomedical informatics has its own jargon (virtually its own language), and I have tried to provide a reasonable definition for terms whose meanings seem to change daily.

0.6. ACKNOWLEDGMENTS

I thank Drs. Adrian Gologan and G. William Moore, who served as primary reviewers. Their careful reading of the first draft convinced me that physicians can understand a book that tethers biology and medicine with computer programming. Their comments greatly strengthened the final work.

Most of all I thank my wife and daughters, who allowed me to digress from my former life as a pathologist to focus on the self-absorbing and less lucrative task of writing science books.

CHAPTER 1

Introduction to Perl Programming (For Absolute Beginners)

1.1. BACKGROUND

One of my former math professors (Gian Carlo Rota) said that every mathematician gets by on three chosen tricks. Once mastered, mathematicians approach all problems in terms of their three tricks. For my own part, most of the software programs I have written are composed of 5 to 25 lines of code and use three little tricks (List 1.1.1). I have used many different programming languages. Whenever I need to program in a new language, I quickly search for the basic language tools that support my tricks, ignoring everything else in the language. I can usually create simple programs within a few hours of tackling an unfamiliar programming language.

With few exceptions, every biomedical computer task can be described as an exercise in these three tricks. Here is an example, written in pseudocode (List 1.1.2). Pseudocode programs are algorithms (the logical steps in a program) expressed in narrative text, rather than in the structured coding syntax of a programming language.

LIST 1.1.1. THREE PROGRAMMING TRICKS IN BIOMEDICINE

File parsing (opening a file and examining the contents of the file, one line at a time)

Pattern matching (finding a fragment of parsed text that matches a word, a phrase, or a character pattern of interest)

Assigning data structures to hold numbers or textual data

LIST 1.1.2. PSEUDOCODE TO COLLECT FROM A FILE ALL THE LINES THAT CONTAIN THE PHRASE "BIOMEDICINE"

1. Open a file for reading. (Verbose equivalent: Get a file from the hard drive that has a particular name and prepare it so that the data in the file can be extracted and put into holders in the computer's memory.)
2. Parse the lines of the file. (Verbose equivalent: Grab the characters from the first line of the file and put that line into a data holder that occupies a specific place in computer memory. Be prepared to repeat this for all the lines of the file.)
3. Collect all the lines that contain the phrase "biomedicine." (Verbose equivalent: As each line is placed in a holder in computer memory, determine whether the line contains the string "biomedicine," and, if it does, add the held data to a structure called an array, which can hold many character strings, in sequence.)
4. When the file is exhausted, empty all the matching lines into an external file that can be viewed in a word processor. (Verbose equivalent: At the end of the file parsing loop, take the array structure, and transfer all the character strings from the array, in sequence, into a newly created file that has been prepared to accept data.)

This pseudocode uses all the tricks that you will need for 90% of your programming tasks: file parsing, pattern matching, and using a data structure (in this case, an array).

In this book, we will not cover many of the programming skills used by advanced programmers (for example, scoping, referencing, debugging, building graphic user interfaces, using pragmas, distinguishing local and global variables, and optimizing code). This book discusses algorithms of importance to biomedical professionals and demonstrates how they might be implemented.

The purpose of this chapter is to provide readers with just enough programming skill to create Perl scripts. Perl scripts are simple, plain-text files.

1.2. DOWNLOADING AND INSTALLING PERL

ActiveState (www.activestate.com) has the most up-to-date Perl versions for many popular operating systems (at the time of this writing). Other sites offering free Perl downloads are *www.perl.com* and *www.cpan.org*. CPAN (Comprehensive Perl Archive Network) is an incredible resource for all PERLish things, including new Perl modules. Perl interpreters for dozens of different operating systems are available from CPAN at: *http://www.cpan.org/ports/*.

Installation procedures can vary greatly. Most installation procedures are easy and have Web-based instruction.

1.3. FILE OPERATIONS

For the biomedical scientist, perhaps the most fundamental programming task is to open a file and display its contents. It is surprising, but virtually all commercial word processors fail at this simple task. If you don't believe this, just try opening a 1 Gbyte text file in your favorite word processor. Biomedical informaticians routinely need to open and read large files composed of sequential plain-text records.

Perl excels at file manipulation. In a few lines of Perl, you can open a file of any size and directly access chunks of data from any location within the file (random file access). Perl lets you open multiple files at once, and lets you iterate through all the files on a drive, or on a network. Perl also lets you extract and analyze data from a text file or a binary file. Perl does these tasks quickly, and with just a few lines of code.

Why are file routines so important? Most biomedical data is accessible in file form. Virtually all genome, proteome, and array datasets are available as simple text files. The data elements are typically separated by commas, by tabs, or (in the case of sentences) by periods (List 1.3.1). We will see how Perl can quickly search through a file of 20 or 30 megabytes, looking for string matches, and returning all of the matching strings, all in about one second.

LIST 1.3.1. SAMPLE CONTENTS OF A TYPICAL FLAT-FILE, "TAXO.TXT," EXTRACTED FROM "TAXONOMY"

```
SYNONYM          : Bacillus aegyptius
SYNONYM          : Haemophilus aegyptius
SYNONYM          : Hemophilus conjunctivitidis
SYNONYM          : Haemophilus influenzae aegyptius
SYNONYM          : Bacillus conjunctivitidis
SYNONYM          : Bacterium aegyptiacum
SYNONYM          : Bacterium conjunctivitis
SYNONYM          : Bacterium pseudo-conjunctivitidis
```

A short Perl script, `open1.pl`, opens a text file and prints the content, line by line (List 1.3.2).

Your first Perl script opens and reads through a text file named `taxon.txt`. The output corresponds to 8 lines from `TAXONOMY.DAT`, a large data file exceeding 36 megabytes (see Appendix). In the next few sections, we will look more closely at our first Perl script.

For the `open1.pl` program to work, there must be a file `taxo.txt` and, in this simple case, the `taxo.txt` file must reside in the same directory as the `open1.pl` Perl script. The output of the file, appearing on the computer screen, consists of a copy of the original `taxo.txt` file. We will fully describe the `open1.pl` script in the next few sections.

LIST 1.3.2. PERL SCRIPT, `open1.pl`, TO OPEN A FILE AND READ A FILE

```
#!/usr/bin/perl
open(FILE, "taxo.txt");
$line = " ";
while ($line ne "")
  {
  $line = <FILE>;
  print $line;
  }
exit;
```

LIST 1.3.3. OUTPUT OF `open1.pl`

```
C:\ftp>perl open1.pl
SYNONYM          : Bacillus aegyptius
SYNONYM          : Haemophilus aegyptius
SYNONYM          : Hemophilus conjunctivitidis
SYNONYM          : Haemophilus influenzae aegyptius
SYNONYM          : Bacillus conjunctivitidis
SYNONYM          : Bacterium aegyptiacum
SYNONYM          : Bacterium conjunctivitis
SYNONYM          : Bacterium pseudo-conjunctivitidis
```

1.4. PERL SCRIPT BASICS

The Perl script (named `open1.pl`) demonstrates the basic structure of every Perl program. First, notice that the Perl program itself is a text file and consists of character text. There is no special programming "environment" in Perl. You can use any text editor to create a Perl file. Notepad will work fine. If you use a word processor, you need to avoid the proprietary mark-up that word processors include in their documents. Make sure that you save the file as a plain-text file. Perl scripts should be given filenames that end with a "`.pl`" file extension. When you want a Perl script to run, you simply call it from the command line. A typical invocation of a Perl script may look like: `c:\>perl open1.pl` (followed by the return key).

In most operating systems, you do not need to type the word "`perl`," because the system knows where the Perl executable is located and also knows that files ending in `.pl` are interpreted by the Perl executable. You might need to enter only the name of the script (followed by the return key), to run your program.

```
c:\>open1.pl
```

The opening line of the Perl script (`#!/usr/bin/perl`), in the UNIX/Linux environment, lists the full directory path to the Perl interpreter. Perl scripts downloaded from the Web might have all sorts of variations of the first line, depending on where the author's Perl script resides (for example, `#!/perl`, `#!/user/local/bin/perl`, `#!/usr/perl`).

If the Perl script does not reside in the same directory as your Perl interpreter, or in your PATH (list of directories your system searches to find Perl), then you may start your program from the Perl interpreter subdirectory and include the full path to the script, for instance:

```
c:\perl>perl c:\scripts\open1.pl
```

If this is not clear to you, please read the next few paragraphs, which are written for non-Linux users who are new to or unfamiliar with command-line instructions executed via a DOS prompt.

1.5. THE DIRECTORY PATH TO PERL

For Linux/UNIX computers, Perl scripts are invoked by typing the name of the Perl script on the command line (the "C" prompt on Windows machines or the shell prompt on Linux/UNIX machines). For non-Linux computers, command lines can be invoked through the MS-DOS shell (available as an application icon labeled "Command Prompt" on Windows systems). A command line is also available through the "Run" dialog box available by first clicking on the "Start" button.

When your Perl interpreter is in the same subdirectory as your Perl script, your script will execute when you invoke "`perl <name of`

`script>`" from the Perl subdirectory command prompt. It might prove impractical to always place your Perl scripts in the same directory in which your Perl interpreter resides. In the DOS system, there is an internally stored list of paths that the operating system searches whenever an executable program is invoked from the command line. You should add the directory in which your Perl interpreter resides to your operating system's path list; this is usually achieved by simply adding the path to Perl in the "path" statement of your system's `autoexec.bat` file. When the directory in which the Perl interpreter resides is listed in your system's path, you can call the Perl interpreter from any directory or subdirectory on your hard drive, and your system will find Perl. In most cases, adding Perl to your system's "path" statement is done automatically for you when you install Perl.

So, if your Perl script (`anything.pl`) is kept in the anywhere subdirectory, then you can execute the script with:

```
c:\anywhere>perl anything.pl
```

1.6. DO NOT PANIC! PERL WILL FORGIVE YOU

Though Perl is a forgiving language, some scripts occasionally "hang" — fail to execute — and leave you waiting for a screen that does not respond to plaintive mouse clicks and pathetic jabs at the return key. Do not worry. Your computer is fine. In Windows systems, pressing the control key and the break key together (control-break) tells Perl to give up and returns you to the command prompt. In Linux, use Control-C. Sometimes it takes a few seconds for Perl to realize that it must stop, but this technique almost always works. If all else fails, control-alt-delete will take you out of the DOS window.

You might notice that when you compose a Perl script, it might fail to compile the first time you try to execute the program. This is okay. Perl begins each execution of a program by interpreting your script. If there are syntax errors, then Perl exits the script and indicates the error. Typically, the Perl error explanation is sufficient for you to go back into the script and correct the problem. It has been my experience that a few types of errors account for the bulk of script failures (List 1.6.1).

Some of these errors will have little meaning to you until you start writing your own Perl programs. You will probably return to this list when you encounter compile errors or anomalous program behavior.

What have we learned so far (List 1.6.2)?

LIST 1.6.1. COMMON ERRORS IN PERL SCRIPTS

Perl blocks must be balanced with curly brackets. Every block (for example, `while`, `if`, `for`, `unless`, `foreach`) must have a beginning curly bracket, "{" and a balanced closing curly bracket, "}". This can become hairy in scripts that have multinested blocks.

Command lines must end with a semicolon.

String variables must be pre-pended with a "$," as in, *`$date`*.

Spelling is important. Perl cannot interpret a misspelled command or variable.

An uppercase character has a different ASCII value from its lowercase equivalent. You will need to maintain case consistency in your Perl scripts.

Characters that serve as reserved Perl symbols must be "escaped" (preceded by a backslash) if they are used as string characters. For example, use \. \/ \\ `\$` if you want to use ./\ or $ as characters. There are exceptions to this rule: `\n`, `\d`, and `\w` are reserved symbols and never refer to the letters, n, d, and w. The strange and nonintuitive use of backslashes in Perl takes some mental adjustment and accounts for the "leaning toothpick syndrome" in Perl scripts. Complex regular expressions often resemble toothpicks tossed amidst string characters.

Certain operations must be enclosed by parentheses (for example, if (1 == 2), not (if 1 == 2)).

The "=" operator assigns a value and does not test for equality. To test for equality, use "==" if you are comparing two numbers and use "`eq`" if you are comparing two strings. Remember that string comparison operators (`eq`, `ne`, `lt`, `gt`) are different from number comparison operators (==, >, <).

Regex operations are preceded by a "=~" operator, not by the assignment operator, "=" (discussed in the next section).

LIST 1.6.2. SUMMARY OF THE FIRST PERL PROGRAMMING SECTION

Perl scripts are simple text files.

Perl scripts should be named using the .pl extension.

Perl is a quintessential command-line language. At the command prompt, run your scripts by typing perl, the name of the script, and then the return key. (On some systems, you needn't include the name perl.)

Perl scripts start off with a header line.

Perl commands end with a semicolon.

Perl blocks are delineated by curly brackets ({ }).

You can assign strings to variables by using the assignment operator, "= ".

You can read, write, or append to files using the "`open`" command.

Most Perl scripts can be represented by a list of imperative Perl commands inside conditional blocks of code that may be deeply nested.

You can add comments to script lines, after a "#" sign.

1.7. PSEUDOCODE FOR A GENERAL-PURPOSE BIOMEDICAL PROGRAM

Virtually every program I write has a common structure (List 1.7.1) and length (10-40 lines). My programs tend to take named files as input and produce named files as the output. The input files can be narrative text, arrays of data, or sequential records. The files are usually found on my computer's hard drive, but I occasionally write programs to parse files scattered across the Internet. With the exception of the actual values of variables, most of my programs can be constructed from bits and pieces of my previously written programs. The generalizability of scripted code and the ability to accomplish most tasks in a few lines of code are compelling reasons to acquire programming skills.

1.8. THE `Open1.pl` SCRIPT, LINE BY LINE

Let us look at the `open1.pl` script (List 1.3.2). The first line of the script has a specific format, consisting of the pound sign, followed by an exclamation point, followed by the directory path leading to the subdirectory in which the Perl interpreter resides. Think of this line as a header that reminds the computer it is reading a Perl script. The meaning and the syntax requirements of the header line are somewhat platform dependent. To make every system happy, be sure to include a header line in each script, and make sure that the header line begins with `#!` (so-called shebang, sharp, or pound sign plus exclamation point) and has the word "perl" somewhere in the line. Many Perl programmers add optional switches to the top line of their Perl scripts that can change the behavior of the Perl compiler and enforce good coding technique.

The script begins with a command to open a file for reading:

```
open(FILE, "taxo.txt");
```

LIST 1.7.1. PSEUDOCODE THAT OUTLINES THE GENERAL CONSTRUCTION OF A PERL SCRIPT.

```
header line;
input something;
if (something evaluates to true)
  {
  do something;
  for or while (some condition)
     {
     do something;
     }
  do something;
  do something;
  }
for or while (some condition)
  {
  do something;
  if (something evaluates to true)
    {
    do something;
    do something;
    do something;
    }
  output something;
  }
exit;
```

This simple command is one of the most powerful and useful Perl commands. When Perl encounters this command, it searches the current directory for the file, named "taxo.txt" in this example. If it finds the file, it assigns the file a filehandle, which is Perl's internal identifier for the file. In this case, the filehandle name is FILE. You should always compose filehandle names in uppercase letters. The file is set to begin reading at the beginning of the file and is ready to accept a variety of file operations, including reading a single line of text (using the <> notation), or moving to different locations in the file. These operations must be called using the assigned filehandle name, not the name of the file as it occurs in its directory. Notice that the line ends with a semicolon. Perl command statements always end with a semicolon. The most common error in writing any Perl script is omitting the obligatory semicolon.

Although Perl provides a variety of parameters that alter the `OPEN` command, examples of the three most useful are the following:

```
open(HANDLE, "some.txt");
```

This opens the named file, "`some.txt`", for reading, assigning the supplied filehandle, "`HANDLE`".

```
open(OUT, ">other.txt");
```

This opens the named file, "`other.txt`", for writing (file writing is determined by the ">" parameter), assigning the supplied filehandle name, "`OUT`," and setting the first write operation for the beginning of the file. Writing to the beginning of a file will overwrite any pre-existing text in the file.

```
open(FUN, ">>third.txt");
```

This opens the named file, "`third.txt`", for appending (file appending is determined by the ">>" parameter), assigning the supplied filehandle name, "`FUN`," and setting the first append operation for the end of the file.

The next line of the script is:

```
$line = " ";
```

This command creates a new string variable (see Glossary). We arbitrarily name the variable "*`$line`*" and assign it the space character (that is, the character created whenever you push the keyboard's spacebar). The space character is flanked by quotation marks. We could have chosen almost any character other than the empty character ("") to initialize the variable. The begin quotes and end quotes are Perl's way of delineating a string. In Perl, the "=" operator assigns whatever is on the right side of the operator to the variable on the left side of the operator. Also, in Perl, every variable begins with the "`$`" sign. The second-most-common error when writing Perl files is to forget to put a "`$`" sign in front of every named variable.

The next line is the following:

```
while ($line ne "")
```

The word "`while`," followed by an expression enclosed by a set of parentheses, constitutes a conditional expression. This tells Perl to evaluate the expression inside the parentheses to determine whether it is true or false. The "`while block`" is entered only if the expression is evaluated to true. Otherwise, Perl skips to the next command following the `while block`. Perl knows where the block starts and ends by looking for the paired curly brackets designating the beginning "`{`" and ending "`}`" of the block. The `while block` will loop forever until the while statement becomes false or until some instruction within the "while block" tells Perl to exit the block.

In later chapters, we will learn several other forms of conditional blocks in Perl, and learn several easy ways to gracefully exit an otherwise endless loop. We will also learn that blocks can be nested (conditional statements within conditional statements). The third-most-common mistake when writing a Perl script is to have unbalanced sets of curly brackets surrounding nested blocks. A consistent programming style (for example, indenting curly brackets and allowing a separate line for each bracket) will reduce such errors.

The expression within the parentheses is the following:

```
$line ne ""
```

Perl has several string comparison operators: ne, eq, lt, and gt. These stand for: not equals, equals, less than, and greater than. These should not be confused with Perl's numeric comparison operators: ==, >, <, >=, and <=. A common mistake for new Perl programmers is to use a string comparison operator (ne, eq, lt, gt) on numeric variables or to use a numeric comparison operator (==, >, <, >=, or <=) on string variables. The use of "ne" tells Perl to determine whether the two string variables flanking the "ne" operator are "not equal" (that is, contain different strings). If they are not equal (that is, their inequality evaluates to true), then Perl proceeds to evaluate the command statements within the block.

The "" represents the empty string (that is, a variable that contains nothing between the quotes). We know that the first test of the while condition will evaluate to "true" because we initiated the variable *$line* by making it nonempty (placing a "space" character into the variable). Because *$line* is not equal to the empty string, Perl will proceed to evaluate the commands in the while block.

The first command statement in the block is the following:

```
$line = <FILE>;
```

The equals sign in Perl is not a logical test for equality (for that, you will use the == sign). Remember, the "=" operator tells Perl to take the value on the right and put it into the variable on the left. In this case, it takes whatever is in <LINE> and puts it into $line. When Perl sees <LINE>, it automatically reads from the file to which the file handle LINE refers, returning one line, the next line from the file. In this case, a line is removed from "taxo.txt" and put into the variable *$line*. Perl has now advanced one line in the file, and is ready to read the next line. When Perl takes a line from the file, the file is not altered in any way. Reading a line from a file is like reading a line from a book.

The next line is the following:

```
print $line;
```

This tells Perl to take the contents of `$line` (containing a line from the file `taxo.txt`) and display it in the default output device, your computer monitor. In the next section, we describe how you can easily instruct Perl to send the variable to another file.

The next line is the end-loop curly bracket ("}"). This marks the logical end of the loop and tells Perl to go back to the loop-initiating line (in this case, the while condition) for evaluation. If the while condition evaluates as true, then the loop is entered again.

In this program, the while loop is entered again and again, as long as there are lines remaining in the file `taxo.txt`. When the file has been entirely read, the block puts the empty string into the *`$line`* variable. At the end of the file, the while loop evaluates as false. Once the file is totally parsed, a line-read returns the empty string, and the statement that `$line` is not equal to the empty string becomes false. When the while condition becomes false, the program bypasses the while loop and goes to the line that follows the loop, in this case, the "`exit;`" line. This tells Perl that the script has finished.

1.9. AN EIGHT-LINE PERL WORD PROCESSOR

Now that we have slogged through the preliminaries, let us try a more ambitious but equally brief Perl script, `mywp.pl`. This short program takes your keyed-in screen input and puts it into two different files, one line at a time (List 1.9.1).

What are the two files that the program creates? One output file is named `mynew.txt`. This file contains only the text that you entered during the current session of the `mywp.pl` script. If there was any text in the `mynew.txt` file when you invoked the `mywp.pl` script, then that text is lost forever.

The other output file is named `mycumu.txt`. This file contains the accumulated input of everything you have ever entered using the `mywp.pl` script, even if you have started the script every day of your life and typed all day each time you used the script, all your typed words over your entire lifetime will be in the `mycumu.txt` file. Try doing that with a word processor!

How does the `mywp.pl` script work? The first two lines of the script create the cumulative file `mycumu.txt`:

LIST 1.9.1. `Mywp.pl`**, A RIDICULOUSLY SHORT TEXT EDITOR, IN PERL**

```
#!/usr/bin/perl
open (OUT, ">>mycumu.txt");
open (NEW, ">mynew.txt");
$line = " ";
until ($line eq "\n")
    {
    $line = <STDIN>;
    print OUT $line;
    print NEW $line;
    }
exit;
```

```
open (OUT, ">>mycumu.txt"');
```

This tells Perl to open a file called `mycumu.txt` as an "append" file, a type of file whose prior contents are not lost when the file is opened for writing. If the file does not already exist, then Perl will create the file for you. The file is assigned the `OUT` file handle.

The next line is the following:

```
open (NEW, ">mynew.txt");
```

This opens another file for writing. Any pre-existing text in the "`mynew.txt`" file will be lost. Perl will create a new file named "`mynew.txt`" if one does not already exist.

With the files created and waiting for your input, the script creates an until loop (List 1.9.2) that will accept keyed input forever or until you press the return key twice.

You are finished! In a few lines of Perl, you have made a word processor. Perl will open the `mycumu.txt` file and have it ready for your input instantly, even if the file is a gigabyte in length. The word processor can be invoked just by typing "`perl mywp.pl`" at the command prompt. It will even keep a current session file for you, in case you only want to review the most recent script session.

The `mywp.pl` script introduces another Perl device, the "#" comment. When a "#" occurs on a command line, Perl ignores the line text that follows. This device allows Perl programmers to add comments within scripts without confusing the Perl interpreter.

LIST 1.9.2. PERL SNIPPET FROM `mywp.pl`, CONTAINING "`until`" LOOP

```
$line = " ";
until ($line eq "\n")  #loop stops when all you have
                           #entered is the return key
  {
  $line = <STDIN>; #waits for the next line of input
  print OUT $line;     #appends to the cumulative file
  print NEW $line;
           #writes to the current script-session file
  }
```

1.10. A SLIGHTLY LONGER COMMAND-LINE TEXT EDITOR

`Edit.pl` is a slightly longer (and much better) text editor (List 1.10.1). This text editor will accept input from the keyboard and prepares two files, one containing your input, and the other formatted so that no line exceeds 70 characters.

The `edit.pl` script introduces several operators that will be strange to the new Perl programmer. We will examine all of these operations in later chapters. For now, let us examine a subroutine contained in the `edit.pl` script.

The `filename_maker` subroutine creates a unique filename so you don't need to think of a new filename for your input text and so that you won't need to worry that the automatically created filename already exists somewhere in the current subdirectory. Remember, write operations to a named file will overwrite any pre-existing data contained in the file, and this could be disastrous to the unwary programmer.

Perl has a simple method for calling subroutines (see Glossary). You simply invoke the subroutine name, prepending an ampersand:

```
&filename_maker;
```

When a subroutine is called, Perl leaves the main body of the script, moves to the subroutine, completes the subroutine, and returns to the line in the script that follows the original invocation of the subroutine.

LIST 1.10.1. PERL SCRIPT `edit.pl`**, A SHORT COMMAND-LINE TEXT EDITOR**

```
#!/usr/local/bin/perl
&filename_maker;
sub filename_maker
  {
  my @listchar;
   for ($count = 1; $count <= 12; $count++)
     {
     push(@listchar, chr(int(rand(26))+65));
     }
   $listchar[8] = ".";
   $filename = join("",@listchar);
   if (-e $filename)
     {
     undef @listchar;
     &filename_maker
     }
   }
open (NEWTEXT, ">$filename");


print "Start typing....Exit program by pressing return key 4
   times\n";
while (1)
  {
  $linesize = <STDIN>;
  print NEWTEXT $linesize;
  if ($linesize eq "\n")
    {
    $total = $total . $linesize;
    if ($total eq "\n\n\n\n")
       {
       print "Your text is saved in file $filename";
       close NEWTEXT;
       last;
       }
    }
  else
    {
    $total = "";
```

```
        }
      }
    open (TEXT, $filename);
    $filename =~ s/[A-Z]$/\_/o;
    #We will use the substitution pattern match to create
    #a new filename by replacing the last character from
    #the original file with the underscore character
    #So, MYFILE.TXT would become MYFILE.TX_
    #The $ sign in the first matching pattern indicates a
    #match evaluated only at the end of the string
    open (NEWTEXT, ">$filename");
    $oldline = "";
    $line = " ";
    while ($line ne "")
          {
          $line = <TEXT>;
          if ($line eq "\#\#\#\n")
            {
            if ($oldline ne "")
                {
                print NEWTEXT "$oldline\n";
                print NEWTEXT $line;
                $oldline = "";
                next;
                }
            }
          my $grepline = $line;
          $grepline =~ s/\n/ /o;
          if ($oldline ne "")
              {
              $grepline = $oldline . " " . $grepline;
              $oldline = "";
              }
          if (length($grepline)>73)
                {
                my @eighty_line=grep{length()>0}split(/(.{1,70})
                +/,$grepline);
                #See Glossary for information on grep()
                $oldline = pop(@eighty_line);
                #we will use pop(), which grabs the last
                #item from an array, in other scripts.
                #see Pop in Glossary
                foreach my $value (@eighty_line)
```

```
                {
                $value =~ s/^ +//o;
                print NEWTEXT  "$value\n";
                }
            }
        else
            {
            $grepline =~ s/^ +//o;
            print NEWTEXT "$grepline\n";
            $oldline = "";
            }
        }
exit;
```

1.11. A SUBROUTINE THAT GENERATES RANDOM FILENAMES

The `filename_maker` subroutine from the `edit.pl` script (List 1.10.1) creates a unique filename consisting of eight characters followed by a period followed by three characters (the standard DOS filename format that can be accommodated by any operating system). The `filename_maker` subroutine follows:

```
sub filename_maker
  {
  my @listchar; #creates an array variable
   for ($count = 1; $count <= 12; $count++)
     {
     push(@listchar, chr(int(rand(26))+65));
     }
   $listchar[8] = ".";
   $filename = join("",@listchar);
   if (-e $filename)
     {
     undef @listchar;
                 &filename_maker
     }
   }
```

Let us see how the subroutine works.

```
for ($count = 1; $count <= 12; $count++)
```

This is a counting "for" loop. We'll visit the counting "for" loop in more detail in the next section, but we can surmise here that Perl starts counting the loop with the number one and continues to loop, incrementing the `$count` variable and halting the loop when `$count` is no longer less than or equal to 12. This produces 12 loops of the enclosed block.

Each looped block executes a single line of code:

```
push(@listchar, chr(int(rand(26))+65));
```

This deceptively simple line of code contains five operations: `push`, `chr()`, `int()`, `rand()`, and `addition`. Perl is a great language for concatenating operations.

Perl starts at the innermost nest (parenthetical operation) and proceeds to the next enclosed parenthetical operation:

```
rand(26)
possible output: 12.8563842773438
```

The `rand()` function returns a random decimal number between 0 and the provided expression. If the provided expression is empty (for example, `rand()`), rand automatically uses 1. The `rand()` function is remarkably useful and we shall re-visit the `rand()` function in Section 14.2.

Because the `rand()` function returns a decimal number, and we will need an integer number, we apply the `int()` operation to the output of the `rand(26)` operation to yield the integer part of an expression (that is, it truncates the decimal portion). This will give us a randomly selected integer that might include 0 and 25 and all integers between these integers.

The number 65 is added to each random character, producing a number from 65 to 90. The `chr()` function converts numbers from 0 to 255 to their ASCII character equivalents. Numbers from 65 to 90 account for uppercase ASCII characters from A to Z. We will re-visit the ASCII table in Section 3.7.

The randomly chosen characters (from A to Z) are pushed into an array using Perl's `push()` operator, which has the syntactic form:

```
push(name of an array, value of an item)
```

The `push()` operator adds items to the end of the array.

The loop produces a 12-item array consisting of randomly chosen uppercase letters.

We want the filename to consist of eight letters followed by a period followed by three additional letters. We substitute a period for character 9 (the eighth array element when the array begins with element number 0), using the array item assignment operator:

```
$listchar[8] = ".";
```

The array is finished. Next, we convert the array to a character string, using the join() function. Join takes an array and concatenates the ordered elements of the array into a character string, with each character separated by a provided joining string. For instance:

```
join("\n", @array)
```

This produces a list of the array elements with each element beginning on a new line.

In our script, the joining character is empty, so all of the elements of the array are concatenated as a simple character string:

```
$filename = join("",@listchar); #"" is the empty character
```

The next lines of the filename_maker script determine whether the filename created already exists in the current subdirectory:

```
if (-e $filename)
  {
  undef @listchar;
  &filename_maker
  }
```

Perl contains a variety of operators to provide information on the status of files. We'll be learning more about these operators in Section 3.2. In this subroutine, we use the exists filetest (-e) to determine if the filename created in the first part of the filename_maker script already exists in the current subdirectory:

```
if (-e $filename)
```

If the file *$filename* already exists, the operator returns a TRUE boolean, and the if block executes.

We undefine the array @listchar (with the undef()) operator. This ensures that the next loop of the push() operator doesn't just add more characters to the pre-existing array of filename characters:

```
undef @listchar;
```

We then send execution back to the beginning of the filename_maker subroutine. If our first passage through the subroutine produced a filename that already existed in the current directory, we will just repeat the routine, hoping to create a unique filename the second time around:

```
&filename_maker
```

This is an example of a recursive subroutine (a subroutine that calls itself) . This subroutine will probably never be called by the script. The assignment of a filename composed of 11 random uppercase letters can produce 26^{11} different filenames. The probability of producing any two identical filenames is essentially zero.

1.12. INTERACTIVELY READING LINES FROM A FILE

Opening large text files in word processors is always a frustrating experience. Depending on your system memory and your word processor, trying to open a large (many megabyte) file can crash or stall your software. Many of the files used in biomedical research exceed 100 megabytes and cannot be opened by word-processor applications.

Perl can read any file and has rapid access to any location in the file. Just use the open command to open a file, and use `seek()` to move to any byte location in the file. Follow those commands with a `read()` command and Perl will promptly export a specified length of text into a variable or file, or display the text on your screen. We will be using `seek()` and `read()` in Chapter 3.4 and Chapter 6.3.

When I download a large file, the first thing I like to do is to read a few hundred lines, starting at the beginning. The `bigread.pl` Perl script will open any file and put the first 20 lines on your monitor. Every time you press the return key, another 20 lines is displayed. You can key through the whole file if you like.

1.13. READING THROUGH LARGE FILES

This script (List 1.13.1 and List 1.13.2) lets you page through enormous files, 20 lines at a time, with no file-load time.

LIST 1.13.1. PERL SCRIPT `bigread.pl`, READS 20-LINE INCREMENTS FROM A LARGE TEXT FILE

```
#!/usr/bin/perl
print "What file do you want to read?";
$filename = <STDIN>;
chomp($filename);
open (TEXT, $filename)||die"Cannot open file";
$line = " ";
while ($line ne "")
    #comment: while $line is not equal to empty
    {
    for ($count = 1; $count <= 20; $count++)
        {
```

```
        $line = <TEXT>;
        print $line;
        }
    print "Type QUIT if you want to quit. Otherwise press any
    key\n";
    $response = <STDIN>;
    if ($response =~ /QUIT/i)
        {
        last;
        }
    }
exit;
```

LIST 1.13.2. PARTIAL OUTPUT OF FILE-READING SCRIPT, `bigread.pl`

```
C:\ftp>perl bigread.pl
What file do you want to read?e:\omim.txt
*RECORD*
*FIELD* NO 100050 *FIELD* TI 100050 AARSKOG SYNDROME
    *FIELD* TX [Grier et al. (1983) reported father and two
    sons with typical Aarskog syndrome, including short
    stature, hypertelorism, and shawl scrotum.
.
.
sons and that this suggested autosomal dominant inheritance.
    Actually, the mother seemed less severely affected,
    compatible with X-linked
Type QUIT if you want to quit. Otherwise press any key
```

How does the bigread.pl script work? The first two lines simply prompt the user for the name of the file to be opened:

```
print "What file do you want to read?";
$filename = <STDIN>;
chomp($filename);
```

The print command implicitly tells Perl to put the argument string into the standard output (`STDOUT`), which is your computer screen.

Look at the next line. `STDIN` is Perl's name for the standard input, your keyboard. This statement tells Perl to wait for input keyed into your monitor. Perl is prepared to wait forever for you to reply to the "What file do you want to read?" prompt. Type in the name of a file, and then press the return key. The command line tells Perl to take whatever is keyed into the monitor and put it into the variable *$filename*. When you press the return key, the text passed to STDIN is automatically appended with a newline character. In many instances, you do not really want the newline character to be attached to your string, so Perl provides a simple way to remove the newline character from the end of a character string (if there is one). It is called chomp().

The next few lines of the script are the following:

```
open (TEXT, $filename)||die"Cannot open file";
$line = " ";
while ($line ne "")
 {
 for ($count = 1; $count <= 20; $count++)
#comment: while $line is not equal to empty
    {
    $line = <TEXT>;
    print $line;
    }
```

Perl opens the file that you provided in response to the prompt (now placed in the variable *$filename*) and assigns it the file handle TEXT. When Perl opens a file, it is ready to start reading from the first byte in the file.

A script might require an external file that is not provided or that cannot be opened because it is being used by another application. In these cases, the programmer will often prefer the script to exit gracefully, rather than thrash onward in a doomed attempt to produce a useful output with inadequate input. The "die" command used in conjunction with the open() function is an example of an error trap (see Glossary).

```
open (TEXT, $filename)||die"Cannot open file";
```

This command tells Perl to exit the script if the file cannot be opened, and to send an informational message, "Cannot open file," to the monitor. The "||" is an "OR" condition. Either Perl successfully opens the file OR Perl dies.

We will make a new variable, *$line*, that will hold the lines from the file. We must be able to read the file line by line until there's nothing left in the file. We will use the "empty" condition for $line as the conditional test that we have not yet reached the end of the file. Remember, the string representation of empty is "", indicating that there is nothing between the beginning and end of the string. Because we will stop reading lines when $line is empty, we must initialize `$line` with a non-empty value. We assign " " (a space character) as the initial value of line.

Next, we begin a nested conditional block. We want to read only 20 lines at a time. We use a counting "`for`" loop to restrict iterations to a set of conditions. We have seen a counting "`for`" loop in the `filename_maker` subroutine. Let's examine the syntax of the "`for`" loop:

```
for (expression1; expression2; expression3)
expression1 is the index variable that marks the first
   iteration (usually 1)
expression2 is the index where the iterations stop
expression3 is the expression that indicates how the index
   will be incremented
```

In the example:

```
for ($count = 1; $count <= 20; $count++)
```

Set the index to one, and increment the value of the index by 1 with each iteration, but stop the loop when the index exceeds 20.

Note that `$count++` is just a Perl shortcut for indicating that the value in `$count` should be incremented by 1, and is equivalent to the assignment:

```
$count = $count +1;
```

So, for 20 loops, the program takes a line of text from the file whose file handle is `TEXT`, and displays that line of text in the standard output (the computer screen).

Needless to say, this loop executes instantaneously in Perl. Once the loop is completed, and 20 lines are shown, the `for` loop finishes and the next line of Perl is interpreted and executed.

The Perl script advances to:

```
print "Type QUIT if you want to quit. Otherwise, press
   any key\n";
   $response = <STDIN>;
   if ($response =~ /QUIT/i)
     {
     last;
     }
   }
exit;
```

A line of text appears on the screen, prompting the user to press a key (to see more text) or to type `QUIT` (followed by the return key, of course) to exit the program.

The line

```
$response = <STDIN>;
```

tells Perl to wait until the user responds to the prompt. When the user types a response, the typed character string is assigned to the *$response* variable.

Then Perl performs a pattern match on the character string containing the user's response:

```
if ($response =~ /QUIT/i)
```

Perl evaluates the `if` expression (the stuff in parentheses) to determine its truth. In Perl, an expression is true if it doesn't evaluate to 0 or false. Had we simply written "`if (5)`", Perl would evaluate the `if` expression as true (because 5 is never 0 or false). In this case, Perl needs to evaluate the pattern `/QUIT/i` to determine whether the pattern matches the variable *$response* (the variable containing the user's reply to the screen prompt).

In Perl, patterns are typically represented as regular expressions delimited by slants:

```
/EXPRESSION/
```

The simplest pattern is a character string. In this case, we use the word "`QUIT`". If the string "`QUIT`" is contained in the variable response, then the "`if`" expression will evaluate to true and the `IF BLOCK` will begin.

Notice that the pattern is followed by the letter i. The i tells Perl that the pattern match is case-insensitive.

```
=~ /QUIT/i,
```

The following items will match the pattern expression:

```
QUIT quit Quit quiT fjsdaklaQuitasdklaf aasquiTasklfafsd
```

All of the preceding items contain `QUIT` (case-insensitive).

In this case, the `IF BLOCK` would instruct the program to end the loop (using the "last" command). When the loop is finished, Perl advances to the next line after the loop, which is the exit command. The exit command, as you might expect, ends the program.

What have we learned in this section (List 1.13.3)?

LIST 1.13.3. SUMMARY OF SECTION 13

Conditional tests
Prompting a user for input
Looping using `for()` and `while()` blocks
Constructing `if()` blocks
Simple pattern matching

1.14. GETTING JUST WHAT YOU WANT WITH PERL REGULAR EXPRESSIONS

Regular expressions add enormous power to Perl scripts (List 1.14.1).

Perl programmers pride themselves on what they can do with a single regular expression (regex, for short). Regular expressions can vary from the simple (for example, matching a specific character string) to the ridiculously obscure. Most Perl language primers cover regular expressions extensively, and at least one book has been devoted exclusively to Perl regular expressions. For our purposes, we will describe the rudiments of regular expressions and provide some of the most useful patterns for matching, substituting, rearranging, and extracting data.

LIST 1.14.1. A FEW THINGS YOU CAN DO WITH A ONE-LINE REGULAR EXPRESSION

Collect the lines from a file that contain a specific word, phrase, number, or character pattern.
Rearrange the content of lines based on matching words, phrases, numbers, or character patterns.
Substitute any alphanumeric character string for any other, for the entire file.

1.15. PSEUDOCODE FOR COMMON USES OF REGEX (REGULAR EXPRESSION PATTERN MATCHING)

Though simple minded, List 1.15.1 describes a scripting process for the most common regex tasks performed by Perl scripts. That is why Perl is sometimes called the "Practical Extraction and Report Language." Once you understand the basic scripting model for parsing files and extracting or substituting text elements based on regular expression matches, then you will have the basic set of Perl skills used in biomedicine.

LIST 1.15.1. PSEUDOCODE USING THE MATCH OPERATOR WITH REGULAR EXPRESSIONS

```
for all the lines of a given file
  {
  put the next line from the file into some variable;
  check the line to see if it matches your regular expression;
  {
  if the line matches the regular expression,
    {
    do something with it, like put it into another file;
    or do an operation on the matching value;
    }
```

LIST 1.15.2. PSEUDOCODE USING THE SUBSTITUTION OPERATOR WITH REGULAR EXPRESSIONS

```
for all the lines of a given file
  {
  put the next line from the file into some variable;
  do a substitution on all of the parts of the line that match
  your regular expression;
  do something with the revised line, like rearranging it;
  then put the rearranged line into another file;
  }
```

1.16. A REGEX EXAMPLE

Let us look at an example script that uses Regex. The example in this section uses a public-domain text from Project Gutenberg. Project Gutenberg is a highly successful effort to create plain-text electronic files of many classic novels. Novels that have outlived their copyright protection fall into the public domain. Edward Gibbons (1737–1794) wrote a fascinating and erudite account of ancient Rome in *The History of the Decline and Fall of the Roman Empire*. This public-domain text can be downloaded at no cost from:

http://www.gutenberg.org/dirs/etext96/1dfre10.txt

For the text example, we use Volume I, with the filename `1DFRE10.TXT`, size 1,656,383 bytes. We will use this file for other scripts, but any large plain-text file will suffice.

In the field of machine translation, narrative text must be parsed into sentences, and the sentences become the grammatic unit that translation software interprets. We will write a simple program that takes narrative text and breaks the text into discrete sentences, assigning each sentence to a line of an output file. To write a sentence parser in just a few lines of Perl, we will utilize everything we have learned about Perl pattern matching.

The following seven-line script (`sentence.pl`, List 1.16.1) parses through the text file "`1DFRE10.TXT`," producing an output file named "`1DFRE10.OUT`". The output file lists the sequentially occurring sentences from the text file on separate lines.

The first command of `sentence.pl` opens a file for reading, and the second command opens a file for writing.

By default, when Perl reads a line of text, it accepts characters until it comes to a newline character, marking the end of a line. Perl allows the programmer to define (or undefine) the character that signifies the end of a line by modifying the built-in record separator, `$/`. By using the

LIST 1.16.1. PERL SCRIPT `sentence.pl`**, A SIMPLE SENTENCE PARSER**

```
#!/usr/local/bin/perl
open (TEXT, "1DFRE10.TXT")||die"Can't open file";
open (OUT,">1DFRE10.OUT")||die"Can't open file";
undef($/);
$string = <TEXT>;
$string =~ s/[\n]+/ /g;
$string =~ s/([^A-Z]+\.[ ]{1,2})([A-Z])/$1\n$2/g;
print OUT $string;
exit;
```

`undef` operator on `$/`, Perl is deprived of an endline indicator, so Perl slurps the entire file into a variable when it comes upon the `<TEXT>` (line assignment) command:

```
undef($/);
#undefines the line delimiter so that Perl reads
#the entire file in one line-read operation
$string = <TEXT>;
#the entire file is put into the variable named $string
```

The next line deploys the substitution operator, which has the following syntax:

```
$string =~ s/<pattern that you match>/<replacement pat-
tern>/option;
```

In this command, every newline character is replaced by a space. The `g` option at the end of the command ensures that Perl will repeat the substitution for every instance of a pattern match:

```
$string =~ s/[\n]+/ /g; #\n is the newline character
```

This ensures that the file contained in `$string` no longer contains any line breaks as every newline character will be replaced by a space.

Next, every occurrence of a succession of characters that are not uppercase letters followed by a period, followed by 1 or 2 spaces, followed by an uppercase letter is replaced by a string consisting of the value of the first matched parenthetical expression, ([^A-Z]+\.[ ]{1,2}), which Perl designates as $1, followed by two newline characters, \n, followed by the second matched parenthetical expression ([A-Z]), which Perl designates as $2:

```
$string =~ s/([^A-Z]+\.[ ]{1,2})([A-Z])/$1\n\n$2/g;
```

This `regex` command is very powerful. It takes the entire file held in the *`$string`* variable (which contains no line breaks) and inserts a line-break character at patterns that are likely to occur at the ends of sentences.

Think about what constitutes a good marker for the end of a sentence by observing the transition between sentence one and sentence two from the previous paragraph:

werful. It tak

You observe that lowercase letters are followed by a period, followed by one space, followed by an uppercase letter. This is the kind of pattern generalized by the regular expression.

This script is not foolproof. But it does a fair job of converting a block of text into a list of sentences.

1.17. REGULAR EXPRESSION MODIFIERS AND EXAMPLES

Virtually all languages that accommodate Regex will offer a variety of Regex modifiers (List 1.17.1) that enhance the versatility of Regex.

LIST 1.17.1. PATTERN MATCH OPTIONS AND DESCRIPTORS

`g`	Match globally (find all occurrences).
`i`	Do case-insensitive pattern matching.
`m`	Treat string as multiple lines.
`o`	Compile pattern only once.
`e`	Evaluate Perl commands in the replacement match of a substitution operation.
`^`	Match the beginning of the line.
`.`	Match any character (except newline).
`$`	Match the end of the line (or before newline at the end).
`( )`	Grouping designates a segment of the matching term that can be assigned to a variable. Strings matching group patterns are assigned variables $1, $2, and $3 sequentially as they appear in the regex expressions.
$1	A variable containing the string that matched the first group in a `regex` pattern.
$2	A variable containing the string that matched the second group in a `regex` pattern.
`[ ]`	Character class, this tells Perl to look for a match to any of the characters included within the square bracket.
`*`	Match 0 (zero) or more times.
`+`	Match 1 or more times.
`?`	Match 1 or 0 (zero) times.
`{n}`	Match exactly *n times.*
`{n,}`	Match at least *n times.*
`{n,m}`	Match at least *n, but not more than m times.*
`\n`	newline.
`\W`	Match a non-word character.
`\w`	Match a word character.
`\s`	Match a whitespace character.
`\S`	Match a non-whitespace character.
`\d`	Match a digit character.
`\D`	Match a non-digit character.

Here are some additional commonly used regular expressions. Readers are encouraged to study these or to include them in short Perl test scripts to see how these regex expressions perform on sample text (List 1.17.2).

LIST 1.17.2. SOME `regex` (REGULAR EXPRESSION) OPERATIONS

`$string =~ s/^ +//o;` Removes leading spaces from a character string.

`$string =~ s/ +$//o;` Removes trailing spaces from a character string.

`$string =~ s/ +/ /g;` Changes all sequences of one or more spaces to just a single space.

`$string =~ s/\n//g;` Gets rid of newline (sometimes called line-break) characters in your string.

`$string =~ s/\b(\w+\.[ ]{1,2})([A-Z])/$1\n$2/g;` Finds the most common sentence delimiter (the end of a word followed by a period followed by one or two spaces, followed by an uppercase letter) and substitutes a newline character so that each new sentence begins on a new line. $1 is the text that matches the first parenthetical expression. $2 is the text that matches the second parenthetical expression.

`$string =~ tr/A-Z/a-z/;` Every uppercase letter is converted to a lowercase letter using the translate operator `(tr/a-z/A-Z/` does the opposite`)`.

`$string = lc($string);` Every uppercase letter is converted to a lowercase letter using the lc operator `(uc($string)` does the opposite).

`$string =~ s/\<[^\<]+\>/ /g;` removes angle-bracketed expressions such as HTML or XML markup.

`$string =~ s/ [a-z]{1}/uc($&)/ge;` changes the first letter of lowercase words to uppercase. The g modifier tells Perl to repeat for each match occurrence, and the e modifier tells Perl to evaluate any code that might be present in the replacement pattern.

We will be using Regex expressions throughout the book. Regex expressions are used in many different programming languages, including Python, Ruby, Java, and XML schema definition language. Regex is difficult to master but well worth the effort.

1.18. COUNTING ALL THE WORDS IN A TEXT FILE

Let us try another Perl script, this time using the `split()` command and revisiting the concept of a Perl array. The script `wc.pl` (List 1.18.1) counts the words in a text file.

The text of *Volume I* of *The History of The Decline and Fall of the Roman Empire* is opened for reading and assigned a handle name, "`TEXT`".

LIST 1.18.1. PERL SCRIPT `wc.pl` **COUNTS THE WORDS IN A FILE IN FIVE COMMANDS**

```
#!/usr/local/bin/perl
open (TEXT, "1DFRE10.TXT");
undef($/);
$all_text = <TEXT>;
@wordarray = split(/[\n\s]+/, $all_text);
print scalar(@wordarray);
exit;
```

```
open (TEXT, "1DFRE10.TXT");
```

The line delimiter is undefined, so that when Perl calls a line, the entire file is slurped into the variable.

```
undef($/);
$all_text = <TEXT>;
```

The next line uses a new command, the `split()` operator:

```
@wordarray = split(/[\n\s]+/, $all_text);
```

Split takes a character string, splits it at each matching instance of a pattern, and puts the fragments into an array. An array is a data structure that contains a list of ordered elements. Perl comes with a variety of useful commands to perform array actions such as accessing the elements of an array, adding elements, deleting elements, and combining elements from different arrays.

To see how the split command works, look at the following example. The first parameter of the split command is a regular expression that matches the letter "i". The second parameter is the word "Mississippi". The word Mississippi is split at every occurrence of the character "i".

```
@array = split(/i/,"Mississippi");
@array becomes a list of ordered items:
M
ss
ss
pp
```

In this case, `@array` is `("M","ss","ss","pp")`. Perl starts counting items at zero, so "M" is the zero-place element of the array named `@array`. Notice that "ss" appears as two different array elements (the first and the second elements of the array).

In wc.pl script, Perl is commanded to split the character string that holds the entire text of *Volume I* of *The History of The Decline and Fall of the Roman Empire*, wherever there are one or more occurrences of a line break (the \n character) or a space.

```
@wordarray = split(/[\n\s]+/, $all_text);
```

This creates an array of all the words in the text. The scalar operation returns the size of the array. The size of the array is simply the number of words in the text.

1.19. FINDING THE FREQUENCY OF OCCURRENCE OF EACH WORD IN A TEXT FILE (ZIPF DISTRIBUTION)

George Kingsley Zipf (1902-1950) gave us Zipf's law, asserting that in a text corpus, the frequency of any word is roughly inversely proportional to its rank in the frequency table (1). This means that the second most frequently occurring word in a text might occur about half as frequently as the first most frequently occurring word in a text.

A practical way of interpreting Zipf's law is that a small amount of words account for most of the occurrences of words in any text. A Zipf distribution is a list of the different words in a text, in the descending order of their occurrence, as shown in the Zipf distribution of this paragraph (List 1.19.1).

LIST 1.19.1. THE ZIPF DISTRIBUTION OF THE PRECEDING PARAGRAPH

```
c:\ftp>perl zipf.pl
00007 of
00005 a
00004 the
00003 words
00003 is
00003 in
00002 zipf
00002 text
00002 occurrence
00002 distribution
00001 zipf's
00001 way
00001 this
00001 their
00001 that
00001 small
```

```
00001 shown
00001 see
00001 practical
00001 paragraph
00001 order
00001 most
00001 listing
00001 list
00001 law
00001 interpreting
00001 for
00001 different
00001 descending
00001 any
00001 amount
00001 account
```

Notice that Zipf's law was not strictly obeyed. The second most frequently occurring word, "a," occurred at nearly the same frequency as the most frequently occurring word. Zipf would surely have insisted that his law is tuned to large texts. A few sentences cannot accurately test Zipf's law.

When the Zipf script parses the entire text of *Volume I* of *The History of The Decline and Fall of the Roman Empire*, the first 10 lines of output are shown (List 1.19.2). Zipf's observation that the ranked frequency of occurrence of words in a text drops inversely to its rank seems to hold.

LIST 1.19.2. THE FIRST 10 ITEMS IN THE ZIPF DISTRIBUTION OF *THE DECLINE AND FALL OF THE ROMAN EMPIRE*

```
26856 the
18032 of
09136 and
06026 to
04654 a
04155 in
03170 was
03081 his
02815 by
02391 that
```

Computational linguists rely on Zipf distributions to identify high-frequency, low-information words that delimit phrases of high information content. Zipf distributions are used to build nomenclatures and indexes. They can be useful for tracking the occurrence of misspelled words. Zipf distributions also can be used as a "signature" for a text and as part of methods to detect plagiarized text or to rank textual concepts.

1.20. A PERL ZIPF SCRIPT

With Perl, a Zipf distribution for text files of any size can be created with just six command lines (List 1.20.1).

This script works very much like our previously described word-counting script. The brunt of the computational work occurs when the text is split into individual words, which are ported into an array variable.

LIST 1.20.1. PERL SCRIPT `zipf.pl` **CREATES A ZIPF DISTRIBUTION IN SIX COMMANDS**

```
#!/usr/local/bin/perl
open (TEXT, "1DFRE10.TXT");
open (OUT, ">1DFRE10.OUT");
undef($/);
$all_text = <TEXT>;
$all_text = lc($all_text);
$all_text =~ s/[^a-z\-\']/ /g;
@wordarray = split(/[\n\s]+/, $all_text);
foreach $thing (@wordarray)
  {
  $freq{$thing}++;
  }
#The Zipf list finished.
#The next lines display the distribution.
while ((my $key, my $value) = each(%freq))
    {
    $value = "00000" . $value;
    $value = substr($value,-5,5);
    push (@termarray, "$value $key")
    }
@finalarray = reverse (sort (@termarray));
print join("\n",@finalarray);
exit;
```

```
$all_text = lc($all_text);
$all_text =~ s/[^a-z\-\']/ /g;
@wordarray = split(/[\n\s]+/, $all_text);
```

Perl has a simple command for converting strings to lowercase: `lc()`, the lowercase operator. It is no surprise that Perl has an uppercase operator, `uc()`, for converting all the characters of a string to uppercase. In this case, the lowercase operator is applied to the entire text held in the variable *$all_text*.

Also, we are interested only in words, not punctuation. The substitution operation, s/[^a-z\-\']/ /g, removes all characters that are not lowercase a through z, a hyphen, or an apostrophe (that is, it removes all characters that are not found in words). Remember that the caret in an expression block forces negation on the expression. So [^a-z\-\'] matches everything that is *not* lowercase a through z, a hyphen, or an apostrophe.

The next three lines of the script count the frequency of occurrence of each word in the array, using a foreach block:

```
foreach $thing (@wordarray)
  {
  $freq{$thing}++;
  }
```

The `foreach` block moves sequentially through an array, item by item for each loop, assigning each encountered item to the

$thing variable for the duration of each loop.

1.21. THE ZIPF SCRIPT INTRODUCES PERL ASSOCIATIVE ARRAYS

The `$freq{$thing}++` introduces a new type of data structure that Perl and many other programming languages use extensively: the associative array. Associative arrays are also known as dictionaries or as hashes (not to be confused with one-way hashes, which will be discussed in Chapter 10).

An associative array consists of an unordered collection of key/value pairs. An example of items in a Perl associative array follows (List 1.21.1).

LIST 1.21.1. EXAMPLE OF AN ASSOCIATIVE ARRAY: `%patient_weight`

```
$patient_weight{"John Public"} = 155;
$patient_weight{"Mary Smith"} = 110;
$patient_weight{"Jules Berman"} = 195;
$patient_weight{"Jules Berman"}++; #evaluates to 196
```

In the example, there is an associative array of key/value, pairs, and the name of the associative array is `%patient_weight`. Perl requires a "%" prefix to designate an associative array (for example, `%patient_weight`). Perl assigns the individual key/value elements of an associate array with a specific syntax (for example, `$patient_weight{"Jules Berman"} = 195;`). "Jules Berman" is the key, and "195" is the value. Once a key/value pair is assigned, a key's value can be accessed (for example, `$patient_weight{"Jules Berman"}` will evaluate to 195). Notice that while the name of the associative array is prefixed with a % sign, the name of a key/value pair contained in the array is prefixed with a $ sign. This is because associative array members are strings.

The last line of the list applies Perl's "increment" operation to the value of the array key:

```
$patient_weight{"Jules Berman"}++;
```

This operation takes the value of `$patient_weight{"Jules Berman"}` and increments it by one, yielding 195 + 1 = 196. Once this operation is completed, the value of `$patient_weight{"Jules Berman"}` is modified, and future invocations of `$patient_weight{"Jules Berman"}` will evaluate as 196.

Returning to the Zipf script, we can see that one line of Perl is sufficient to transform an array of items (some repeated) into an associative array wherein the value of each key is the number of times that the key occurs in the original array:

```
foreach $thing (@wordarray)
  {
   $freq{$thing}++; #increments the value of the key/value
        #element $freq{$thing} by 1. Perl creates
        #an associate array %freq even though you
        #haven't declared the associate array.
  }
```

As each array item is parsed, it is assigned to the variable *$thing*, and looped through the list of instructions for the loop block. In this case, the array being parsed is the sequence of words appearing in *Volume I* of *The History of The Decline and Fall of the Roman Empire.*

The loop itself contains just one instruction:

```
$freq{$thing}++;
```

This line creates an associative array element, with a key of the string variable *$thing* (corresponding to an occurrence of a word in the text) and a value corresponding to an incremented count of the number of times that the word has been looped through the block. If the loop encounters a word for the first time, Perl implicitly increments from zero, and assigns a value of one to the associative array key corresponding to the word. The next time the word is encountered in the loop block, the value will be incremented by one. In this way, every word in the text is assigned an item in the `%freq` associative array, with the word serving as the item key, and the number of occurrences of the word as the item value.

After every word in the array of words from the text has been looped through the block, the associative array is complete.

1.22. FORMATTING OUTPUT IN THE ZIPF SCRIPT

The next lines simply format and print the items from the `%freq` associative array, producing a Zipf distribution in descending order of word frequency in the text:

```
while ((my $key, my $value) = each(%freq))
    {
    $value = "00000" . $value;
    $value = substr($value,-5,5);
    push (@termarray, "$value $key")
    }
@finalarray = reverse (sort (@termarray));
print join("\n",@finalarray);
```

A `while` block is created to loop through each member of the `%freq` associative array. This loop command has many applications, and its syntax should be studied:

```
while ((my $key, my $value) = each(%freq))
```

Each loop of the `while` block pulls one key/value pair from the `%freq` associative array. The `while` block continues until the contents of the associative array are exhausted.

A string of five 0s is prefixed to each value in the `%freq` associative array, using the Perl string concatenation operator, "."

```
$value = "00000" . $value;
```

This reformats the number of occurrences of each word in the text. For example, "33" would become "0000033".

The next line uses Perl's `substr()` operator to extract the last five characters from `$value` (see Glossary for more on the `substr()` operator). This ensures that every value consists of a total of exactly five digits front-justified with 0s:

```
$value = substr($value,-5,5);
```

So, "0000033" becomes "00033".

As each transformed value is created, it is immediately added to a new array (named `@termarray`) using Perl's `push()` command. The `push()` command takes a variable and adds it to the end of an array, as shown:

```
push (@termarray, "$value $key")
```

By turning every value in the `%freq` array into a number that is exactly five digits in length, and front-justified with 0s, the newly created array of values can be sorted using Perl's sort function.

In the next line, the five-digit values corresponding to the number of occurrences of words in the text file are sorted, then reversed, to provide a descending array of word frequencies:

```
@finalarray = reverse (sort (@termarray));
```

This reverse-sorted array is then printed to the monitor. The `join()` operator creates a string consisting of each member of the array joined by a chosen string (in this case, the newline character). This produces a list wherein each member of the array appears at the beginning of a new line:

```
print join("\n",@finalarray);
```

What have we learned in Sections 1.14 through 1.22 (List 1.22.1)?

LIST 1.22.1. SUMMARY OF SECTIONS 1.14–1.22

Creating and interpreting complex regular expressions
Looping through arrays with `foreach` blocks
Looping through associative arrays with `while` blocks
New Perl operators and commands: `split()`, `push()`, `lc()`, `sort()`, `join()`, `substr()`, `scalar()`, and `undef()`
Incrementing values and concatenating strings
Advanced pattern substitution and substitution options

1.23. CREATING A PERSISTENT DATABASE OBJECT

Until now, we have studied scripts that take text files as input and create text files as output. Ordinarily, data objects (string variables, arrays, and associative arrays) created in a Perl script will vanish once the script has finished executing.

Perl provides a mechanism whereby data objects can persist, even after the script has finished executing. This is accomplished with Perl's built-in SDBM module. The SDBM module builds an external database that contains the data structures created in your scripts. Any script can access these persistent data structures by simply calling the external database. You can copy and distribute the persistent database object just like any other file, and you can write specialized Perl scripts that retrieve or modify data held by the persistent object. This database functionality in Perl is particularly useful when your scripts create millions of data records.

The following Perl script creates an external database file from the MESH (see Appendix) flat-file (List 1.23.1 and List 1.23.2).

LIST 1.23.1.A SAMPLE MESH (MEDICAL SUBJECT HEADING) RECORD

```
*NEWRECORD
RECTYPE = D
MH = Heparin
AQ = AA AD AE AG AN BI BL CF CH CL CS CT DF DU EC GE HI
   IM IP ME PD PH] PK PO RE SD SE ST TO TU UL UR
PRINT ENTRY = Heparinic Acid|T118|T121|T123|
   NON|EQV|UNK (19XX)|800523|abbbcdef
PRINT ENTRY = alpha-Heparin|T118|T121|T123|NON|NRW|
   UNK (19XX)|800523|abbbcdef
ENTRY = Liquaemin|T118|T121|T123|TRD|NRW|UNK
   (19XX)|861029|abbbcdef
ENTRY = Sodium Heparin|T118|T121|NON|NRW|UNK
   (19XX)|830330|abbcdef
ENTRY = Heparin, Sodium
ENTRY = alpha Heparin
MN = D09.698.373.400
PA = Anticoagulants
PA = Fibrinolytic Agents
EC = antagonists & inhibitors: Heparin Antagonists
MH_TH = BAN (19XX)
ST = T118
ST = T121
ST = T123
N1 = Heparin
RN = 9005-49-6
```

MS = A highly acidic mucopolysaccharide formed of equal] parts of sulfated D-glucosamine and D-glucuronic acid with] sulfaminic bridges. The molecular weight ranges from 6six to] twenty thousand. Heparin occurs in and is obtained from, for example, liver,] lung, and mast cells of vertebrates. Its function is unknown,] but it is used to prevent blood clotting in vivo and vitro, in the form of many different salts.

PM = /therapeutic use was HEPARIN, THERAPEUTIC 1965

HN = /therapeutic use was HEPARIN, THERAPEUTIC 1965

MED = *1635

MED = 3275

M90 = *2406

M94 = 4517

MR = 20040707

DA = 19990101

DC = 1

UI = D006493

LIST 1.23.2. PERL SCRIPT `persist.pl` CREATES A PERSISTENT DATABASE OBJECT FROM THE `MESH` FLAT FILE

```
#!/usr/bin/perl
use Fcntl;
use SDBM_File;
tie%item, "SDBM_File", 'mesh', O_RDWR|O_CREAT|O_EXCL, 0644;
untie%item;
open (TEXT, "d2006.bin")||die"Can't open file";
$/ = "*NEWRECORD";
$line = " ";
while ($line ne "")
    {
    tie%item, "SDBM_File", 'mesh', O_RDWR, 0644;
    #use the created persistent database object
   $line = <TEXT>;
    @linearray = split(/\n/,$line);
    foreach $piece (@linearray)
      {
      if ($piece =~ /MN = /)
        {
        $meshno = $';
        }
      if ($piece =~ /ENTRY = /)
```

```
          {
          $entry = $';
          if ($entry =~ /\|/o)
             {
             $entry = $`;
             }
          $entry =~ s/s\b//g;
          $entry = lc($entry);
          push (@synonyms, $entry);
          }
        }
      foreach $term (@synonyms)
        {
        $item{$term} = $meshno;
        }
     undef $meshno;
     undef @synonyms;
     untie%item;
   }
undef(%item);
close TEXT;
exit;
```

The first lines of `persist.pl` tie a Perl associative array to a persistent database object:

```
use Fcntl;
```

This command tells Perl to use the built-in Perl module `Fcntl` (file control), because you will create a file using the `Fcntl` syntax:

```
use SDBM_File;
```

This command tells Perl to use the built-in Perl module `SDBM_File`, because you will be creating a database object:

```
tie %item, "SDBM_File", 'mesh', O_RDWR|O_CREAT|O_EXCL, 0644;
```

This line does not need to be explained. It is simply Perl's special syntax for creating an external database object file (mesh, in this example) and "tying" it to the associative array `%item`. From this point on, you can use the associative array, `%item`, just like any Perl associative array, but the data structure will be automatically encapsulated in the external file, mesh. Actually, Perl creates two files, `mesh.pag` and `mesh.dir`. Both of these files are binary files. They can be accessed for read/write operations by tying them back to an associative array (next script):

```
untie%item;
```

You can untie the associative array from the database object, and you can tie them back together again when you need to read or write data.

```
open (TEXT, "d2006.bin")||die"Can't open file";
```

`d2006.bin` is the MESH (see Appendix) flat file. We could have chosen any file. We will parse data elements from the MESH flat file and assign them to the `%item` associative array.

```
$/ = "*NEWRECORD";
```

The records in the MESH `d2006.bin` file are multi-line and each record begins with the "`*NEWRECORD`" string. By assigning "`*NEWRECORD`" as the line delimiter, each line read will place an entire multi-line MESH record into a variable.

The remainder of the Perl script is straightforward, and we will discuss only the first few lines of the `while` loop:

```
while ($line ne "")
   {
   tie%item, "SDBM_File", 'mesh', O_RDWR, 0644;
   #use the created file
   $line = <TEXT>;
   @linearray = split(/\n/,$line);

   tie %item, "SDBM_File", 'mesh', O_RDWR, 0644;
```

This line once more ties the %item associative array to the already created persistent database, named mesh. The persistent database is set to read/write access:

```
   $line = <TEXT>;
   @linearray = split(/\n/,$line);
```

During each loop, one record in the MESH flat file is moved into the *`$line`* variable. Then `$line` is split into an array, where each item in the array is a line of the multi-line MESH record.

The next lines of the script merely parse the record into data elements. The important step comes when the `%item` associative array is built:

```
$item{$term} = $meshno;
```

As each key and value is added to the `%item` associative array, the tied persistent database object is likewise built. Finally, after all the parsed MESH elements are added to the associative array, the persistent database object can be untied:

```
untie%item;
```

The purpose of the script was to build a persistent external database object. When the script ends, all non-persistent items vanish into the bit void. This includes string variables, such as `$line`, and associative array variables, such as `%item`. But the persistent database object now exists as two portable files (`mesh.pag` and `mesh.dir`) containing the MESH terms and numbers that can be directly accessed without rebuilding their data structures. In the next script, we show how that object can be used to retrieve data.

1.24. RETRIEVING INFORMATION FROM A PERSISTENT DATABASE OBJECT

Once built, accessing the data in a persistent database object is easy and fast (List 1.24.1).

LIST 1.24.1. PERL SCRIPT `retrieve.pl` RETRIEVES A PERSISTENT DATABASE OBJECT FROM THE `MESH` FLAT FILE

```
#!/usr/bin/perl
use Fcntl;
use SDBM_File;
tie%item, "SDBM_File", 'mesh', O_RDWR, 0644;
while(($key, $value) = each (%item))
  {
  print "$key => $value\n";
  }
untie%item;
exit;
```

In `retrieve.pl`, we use the now-familiar syntax to tie the created data file "mesh" to a Perl associative array, `%item`:

```
use Fcntl;
use SDBM_File;
tie%item, "SDBM_File", 'mesh', O_RDWR, 0644;
```

Perl looks for "mesh" in the external files `mesh.pag` and `mesh.dir` in the current directory. Once tied, `%item` becomes the same as the external mesh database object (that is, the files `mesh.pag` and `mesh.dir`):

```
while(($key, $value) = each (%item))
  {
  print "$key => $value\n";
  }
untie%item;
exit;
```

We describe the `while(($key, $value)` loop back in Section 1.20. The while`(($key, $value)` loop collects all of the key value pairs in the `%item` associative array.

1.25. VALIDATING XML TAGS USING REGULAR EXPRESSIONS

It is rare to find a Perl script that has no regular expressions. Almost all script input (even lists of numbers) need to be validated. Is 1,000,000 acceptable when your mathematical operator expects 1000000? Is .5 acceptable when your script expects all numbers to begin with an integer followed by a decimal? In many cases, passed data elements must be validated against a regex expression that enforces a particular data format.

We will be learning a lot about XML in a later chapter. For now, suffice it to state that XML tags are angle-bracketed descriptors of the information that exists between the start and end of the tag. For example: `<name>Jules Berman</name>`

Let us look at two examples where Perl regular expressions are used to validate XML markup tags and parse free-text sentences.

XML tags (also known as markup) must satisfy structural requirements (List 1.25.1 and List 1.25.2).

Reviewing the requirements for a valid XML tag, we can see that the `tagcheck.pl` script correctly classified tags in every instance. Can you see why the script rejected the invalid terms?

Let us look at how the `tagcheck.pl` script works. The first step in the script is to prepare an array of would-be gene tags:

```
@elements = qw (gene 4gene gene:ncbi gene-autry ge::ne
                   gene&autry -gene _gene gene- gene:
                   :gene ge:n:e  ge:ne:  ge,ne ge.ne);
```

LIST 1.25.1. SYNTAX RULES FOR VALID XML TAGS

XML tags, like Perl variables, are case-sensitive. ("Name" is different from "name".) Parsers must preserve character case.

Letters, underscores, hyphens, periods, and numbers may be used in a tag.

Only letters and underscores are eligible as the first character.

Colons are allowed, but only as part of a declared namespace prefix. For all practical purposes, this means that only one colon is allowed in a tag, and the colon must appear in an internal location in the tag (not at the beginning or the end of a tag).

LIST 1.25.2. PERL SCRIPT `tagcheck.pl` VALIDATES XML TAGS

```
#!/usr/bin/perl
@elements = qw (gene 4gene gene:ncbi gene-autry ge::ne
                gene&autry -gene _gene gene- gene:
                :gene ge:n:e  ge:ne: ge,ne ge.ne);
foreach $value (@elements)
{
if ($value =~/^[a-z\_][a-z0-9\-\.\_]*[\:]?[a-z0-9\-\.\_]*$/i)
    {
    print "$value is good\n";
    }
  else
    {
    print "$value is bad\n";
    }
}
exit;
```

The `qw` operator is a handy device that accepts array lists without commas or string quotes. This operator was made for lazy Perl programmers who do not want to worry about properly "commifying" a list.

Each member of the array is examined (in a `foreach` loop) for a pattern match to a regular expression that embodies every criterion for a proper XML tag:

```
if ($value =~ /^[a-z\_][a-z0-9\-\.\_]*[\:]?[a-z0-9\-\.\_]*$/i)
```

What does this regular expression do? It requires that the string begin with one letter between a and z or an underscore. The ^ symbol appearing in the beginning of a pattern expression indicates the beginning of the string. A block "[...]" indicates a permitted match to any of the characters listed in the block. The first character of a proper XML tag must be a letter between a and z or the underscore character. The characters following the first letter must be a letter between a and z, a number character (0-9), a hyphen, an underline character, or a period. It may also be the colon prefix, but if it is a colon prefix, then it can occur only once. The colon prefix is put into its own block. The "?" after a block indicates that the contents of the block (the colon character, in this case) can occur zero time or one time. Following the colon, if it occurs, can be any number of occurrences of a letter of the alphabet, an underscore, a period, or a hyphen. At the end of the pattern holder (/...../) is the letter "I," indicating that the match pattern can be either uppercase or lowercase letters of the alphabet. Review the output from the `tagcheck.pl` script to show yourself that it really works (List 1.25.3).

LIST 1.25.3. OUTPUT OF `tagcheck.pl`

```
c:\ftp>perl tagcheck.pl
gene is good
4gene is bad
gene:ncbi is good
gene-autry is good
ge::ne is bad
gene&autry is bad
-gene is bad
_gene is good
gene- is good
gene: is good
:gene is bad
ge:n:e is bad
ge:ne: is bad
ge,ne is bad
ge.ne is good
```

1.26. COLLECTING ITEMS PRESENT ON EACH OF TWO LISTS

This script parses the items on two lists and prints a list of items that are present on both lists. Perl has many different ways of achieving the same functionality (List 1.26.1).

This script uses nested blocks to iterate through the items of two lists, checking equality in all possible combinations of two items, each of which are selected from one of the lists.

```
foreach $item (@wordlist)
  {
  foreach $thing (@takeout)
    {
```

Nesting foreach statements, as shown, directs Perl to take the first item from an array (`@wordlist`) and then take, in sequence and separately, every item from another array (`@takeout`) and perform the commands in the block. Then, Perl takes the next item from `@wordlist` and does it all over again, with each and every item from `@takeout`. Perl repeats the long exercise with every item from `@wordlist`.

Perl will nest deeper loops, if you like, iterating through every desired combination of items from the three lists.

LIST 1.26.1. PERL SCRIPT `double.pl` FINDS ITEMS PRESENT ON EACH OF TWO LISTS

```
#!/usr/local/bin/perl
$holder = "";
$takeout =
   "the,end,many,for,not,where,forever,friday,postage,why,can,not";
@takeout = split(/,/,$takeout);
$wordlist =
   "carry,wherever,end,try,stand,postage,many,yesterday";
@wordlist = split(/,/,$wordlist);
foreach $item (@wordlist)
  {
  foreach $thing (@takeout)
    {
    if  ($item eq $thing)
      {
      $holder = $holder . " " . $item;
      }
    }
  }
print $holder;
exit;
```

```
foreach $item (@wordlist)
  {
  foreach $thing (@takeout)
    {
      foreach $item (@yet_another_list)
          {
```

1.27. COLLECTING ITEMS PRESENT ON ONE LIST AND ABSENT FROM ANOTHER

Here is a simple script that finds the words present in a first list and absent from a second list. The script assumes that each file consists of a list of words or phrases, with one word or phrase on each line (List 1.27.1).

This script uses the `exists()` function, which returns a true Boolean only if the parenthetical hash key exists in the hash.

The ability to collect combinations of items from multiple arrays and perform an arithmetic or string operation on the combinations of items will be used again and again in computational biomedicine.

LIST 1.27.1. PERL SCRIPT `differ.pl` FINDS WORDS PRESENT IN ONE LIST AND ABSENT IN ANOTHER

```
#!/usr/local/bin/perl
open(TEXT,"umlsword.txt");
#Let us say that list one is the collection of all
#the different words contained in the UMLS methathesaurus
$inputline = " ";
while ($inputline ne "")
   {
   $inputline = <TEXT>;
   chomp($inputline); #discards the newline character
   $wordlist{$inputline} = 1;
   #assigns an arbitrary value to the %wordlist keys
   }
open(OTHER,"lhodout.txt");
#expects another wordlist file named lhodout.txt in this case
open(DIFFER,">differ.txt");
#puts the missing words in differ.txt
open(SAME,">same.txt"); #puts the common words in same.txt
$inputline = " ";
while ($inputline ne "")
   {
   $inputline = <OTHER>;
   chomp($inputline);
   unless (exists($wordlist{$inputline}))
      {
      print DIFFER "$inputline\n";
      #word or phrase is absent from the first file
      #so we'll put it in the DIFFER file
      print "$inputline\n";
      }
   else
      {
      print SAME "$inputline\n";
      }
   }
end;
```

1.28. WHAT HAVE WE LEARNED (LIST 1.28.1)?

LIST 1.28.1. WHAT WE HAVE LEARNED SO FAR?

The =~ operator tells Perl to look for the pattern that follows the operator in the variable that precedes the operator. Regular expressions are Perl's way of describing a pattern.

You can create most of your patterns by following a few simple rules and by "borrowing" regular expressions from published listings.

The most common usage for regular expressions is in scripts that examine a line (or all the lines) from a file, and that perform a substitution, rearrangement, or other operation on the line, based on the results of the pattern match.

Regular expressions are a powerful and fast tool for modifying text or data records or finding exactly what you want in any text.

Perl associative arrays can be tied to an external database object that persists even when the Perl script has finished executing.

CHAPTER

2

Using Perl Modules and Applying Some Fundamental Commands

2.1. BACKGROUND

We used a few Perl modules (`sdbm` and `fcntl`) in Chapter One. Perl comes bundled with hundreds of useful modules that greatly extend the functionality of the Perl language. A Perl module is a special script that has the suffix .PM, and which is constructed to provide methods that operate in a "protected" namespace (that is, an environment wherein variable names, objects, and data will not conflict with anything you write in your own script). In addition to modules that are bundled into the standard Perl distribution, there are literally thousands of specialized Perl modules available for download on the Internet.

The largest archive of Perl modules is the Comprehensive Perl Archive Network:

http://www.cpan.org

Another wonderful resource is the ActiveState PPM (Perl Package Manager), which is available to anyone who installs the free, ActiveState Perl distribution:

http://www.activestate.com

Most of the modules that you will use come preinstalled in the standard Perl distribution. Installing external modules is usually simple and examples of installations will be described throughout this chapter.

It is easy to use an installed Perl module (List 2.1.1).

We will be using bundled and imported Perl modules throughout this book.

The purpose of this chapter is to show readers how easy it is to find freely available Perl modules that can be easily implemented from their own Perl scripts.

LIST 2.1.1. GENERAL INSTRUCTIONS FOR USING ANY INSTALLED PERL MODULE

1. Read the "usage" section of the module to discover the methods included in the module.
2. Fsrom your Perl script, invoke the module using the "`use <module name>;`" command-line.
3. Use the methods from the module much the same way you would use a built-in Perl command, employing the syntax specified in the "usage" section of the module.

2.2. SORTING BIG FILES

The four-line script `filesort.pl` is one of the most useful Perl scripts in my repertoire. I often have lengthy lists of items (hundreds of thousands) that need to be alphabetically sorted. Nomenclatures, concordances, indexes, anything that must be searched by humans, all need to be sorted. In Chapter 6.4, we will learn that sorted files (that is, files where each line of the file is alphabetically sorted) can be rapidly searched for individual items (or records) using a fast binary search.

Perl has its own sort function, but it is basically intended to sort relatively small arrays. It will not sort large files. For heavy-duty sorting, I use the File-Sort module, which is easily downloaded from the activestate ppm download library or from CPAN.

On my computer, which has a modest 512 Megabytes of RAM, this module sorts files that are hundreds of megabytes in length (List 2.2.1).

LIST 2.2.1. PERL SCRIPT `filesort.pl`, SORTS A LARGE FILE

```
#!/usr/bin/perl
use File::Sort qw(sort_file);
open(STDOUT,">perlbig.out");
#the sorted file will be put
#into perlbig.out
$filename = "perlbig";
#perlbig is the sample file for this script
sort_file($filename);
exit;
```

2.3. A PERL MODULE FOR BENCHMARKING

Programmers today seem less concerned about execution speed than programmers of just a decade ago. This is due entirely to the increased speed and greater RAM (volatile memory) of modern computers. The perceived unimportance of speed is another factor that enhances the popularity of interpreted languages, such as Perl, Python, and Ruby. However, there are times when a script's execution speed will become important to you, particularly if you are parsing large files or performing any kind of natural language parsing (see Chapter 8)

Perl includes a module called `Benchmark.pm` (List 2.3.1).

Does the code in the `bench.pl` script conform to the style of Perl coding used in previous scripts? Not particularly. The code in the `bench.pl` script was essentially copied from the usage text included in the `Benchmark.PM` module. It is often the case that a module will specify a syntax for employing the module. The programmer simply adapts the specified syntax to fit her data.

LIST 2.3.1. PERL SCRIPT `bench.pl` USES THE BENCHMARKING MODULE TO TIME SCRIPT OPERATIONS

```
#!/usr/bin/perl
use Benchmark;
      timethese(1000000,
      {
      'regex1' => '$str="ABCDEFG"; $str =~ s/^(ABC)//; $ch = $1',
      'regex2' => '$str="ABCDEFG"; $str =~ s/^(BC)//; $ch = $&',
      'substr' => '$str="ABCDEFG"; $str=substr($str,3,1)'
      });
exit;
```

The results will be returned like this:

```
c:\ftp>PERL bench.pl
Benchmark: timing 1000000 iterations of regex1, regex2, substr...
regex1:  2 wallclock secs ( 1.81 usr +  0.00 sys =  1.81 CPU) @ 551571.98/s
(n=1000000)
regex2:  1 wallclock secs ( 0.69 usr +  0.00 sys =  0.69 CPU) @ 1453488.37/s
(n=1000000)
substr: -1 wallclock secs ( 0.44 usr +  0.00 sys =  0.44 CPU) @ 2288329.52/s
(n=1000000)
```

2.4. A PERL MODULE FOR TIMING SUBROUTINES

Subroutines are often called upon to do repetitive computational tasks and can account for the bulk of execution time in a Perl script. Often, finding slow subroutines and improving their performance can improve the speed of a Perl script.

Perl comes bundled with an excellent profiling package, `Devel::DProf`, which will time subroutines within a Perl script.

The DProf profile works directly from the command line. I have a Perl script, `neoself.pl`, which calls several subroutines and a module that has dependencies on other modules.

Here is an example of how DProf works with `neoself.pl`:

```
c:\ftp>perl -d:DProf neoself.pl
```

The `neoself.pl` script executes just as it would without the inserted command "`-d:DProf`". When it finishes executing, I input:

```
c:\ftp>dprofpp
```

The following output appears in the DOS window (List 2.4.1).

LIST 2.4.1. USING THE PERL CODE PROFILER `DProf`

```
Total Elapsed Time = -1.70421 Seconds
  User+System Time = 46.81672 Seconds
Exclusive Times
%Time ExclSec CumulS #Calls sec/call Csec/c  Name
13.2   6.195  5.983 132954   0.0000 0.0000
   main::handle_elem_end
7.01   3.283 12.521      1   3.2833 12.520
   XML::Parser::Expat::ParseStream
4.75   2.225  2.012 132954   0.0000 0.0000
   main::handle_elem_start
3.53   1.653  1.227 266050   0.0000 0.0000
   main::handle_char_data
0.07   0.032  0.063     10   0.0032 0.0063
   LWP::UserAgent::BEGIN
 0.07   0.031  0.140      4   0.0077 0.0351  XML::Parser::BEGIN
 0.03   0.016  0.016      1   0.0160 0.0160  warnings::BEGIN
0.03   0.016  0.016      4   0.0040 0.0040
   HTML::HeadParser::BEGIN
0.03   0.016  0.031      4   0.0040 0.0077
   XML::Parser::Expat::BEGIN
```

```
  0.03    0.015   0.015     303    0.0000 0.0000   IO::Handle::read
  0.03    0.015   0.015       6    0.0025 0.0025   IO::Handle::BEGIN
0.03    0.015   0.015       4    0.0037 0.0037
   URI::WithBase::BEGIN
  0.03    0.015   0.155       1    0.0150 0.1552   main::BEGIN
  0.00    0.000  -0.000       2    0.0000       -   SelectSaver::BEGIN
  0.00    0.000  -0.000       2    0.0000       -   IO::BEGIN
```

DProf is an excellent tool for examining the flow of a script through its subroutines. It is another example of the utility of bundled Perl modules.

2.5. APPROXIMATE WORD MATCHING WITH THE String::approx MODULE

Approximate word matching is a wonderful tool, particularly in biomedicine, which contains a rich vocabulary of hard-to-spell terms. Perl provides the `String::Approx` module that accepts input words and finds similar words from a provided word list (List 2.5.1 and List 2.5.2).

The script required an external file containing vocabulary words. In this case, the script looked for the file named "`meshword.txt`" in the `c:\ftp\subdirectory`. This list contains all the different words included in the National Library of Medicine's MESH (Medical Subject Headings) nomenclature (see Appendix).

LIST 2.5.1. PERL SCRIPT `approx.pl`, FOR CREATING A LIST OF WORDS SIMILAR TO A QUERY WORD

```
#!/usr/local/bin/perl
use String::Approx qw(amatch);
print "What word would you like to approximate?\n";
$givenword = <STDIN>;
print "\n";
open(DICT, "meshword.txt") ||die"Cannot";
$line = " ";
while($line ne "")
   {
   $line = <DICT>;
   print "similar to\: $line" if (amatch($givenword, $line));
   }
close DICT;
exit;
```

LIST 2.5.2. OUTPUT OF `approx.pl`

```
c:\ftp>perl approx.pl
What word would you like to approximate?
hematomato

similar to: hematomas
similar to: hematoma

c:\ftp>perl approx.pl
What word would you like to approximate?
hemocromatosis

similar to: hemochromatoses
similar to: hemochromatosis

c:\ftp>perl approx.pl
What word would you like to approximate?
canser

similar to: canker
similar to: anticancer
similar to: scanner
similar to: ganser
similar to: cancer

c:\ftp>perl approx.pl
What word would you like to approximate?
malilgnant

similar to: premalignant
similar to: malignant
similar to: nonmalignant

c:\ftp>perl approx.pl
What word would you like to approximate?
collagin

similar to: collagen
similar to: protocollagen
similar to: tropocollagen
similar to: procollagen
```

```
c:\ftp>perl approx.pl
What word would you like to approximate?
illium

similar to: pseudomycellium
similar to: verticillium
similar to: tiglium
similar to: lilium
similar to: acidiphilium
similar to: spirillum
similar to: thallium
similar to: allium
similar to: sagittifollium
similar to: vitallium
similar to: william
similar to: rhodospirillum
similar to: eupenicillium
similar to: magnetospirillum
similar to: motilium
similar to: trillium
similar to: anaerobiospirillum
similar to: methanospirillum
similar to: dipyridilium
similar to: cilium
similar to: gastrospirillum
similar to: auxilium
similar to: beryllium
similar to: ilium
similar to: penicillium
similar to: psyllium
similar to: tranxilium
similar to: gallium
similar to: azospirillum
similar to: herbaspirillum
```

2.6. CREATING A WORD LIST

It is easy to generate a list of correctly spelled medical words (List 2.6.1):

`meshword.pl` produces the file meshword.txt, consisting of 64,690 different words, one word per line. A sample from the output file demonstrates the rich vocabulary included in MESH (List 2.6.2).

LIST 2.6.1. PERL SCRIPT `meshword.pl` **EXTRACTS A WORDLIST FROM MESH**

```
#!/usr/local/bin/perl
open (TEXT, "d2006.bin")||die"cannot";
open (OUT, ">meshword.txt")||die"cannot";
$line = " ";
while ($line ne "")
  {
  $line = <TEXT>;
  $getline = $line;
  $getline = lc($getline);
  $getline =~ s/[^a-z]/ /g;
  @linearray = split(/ +/,$getline);
  foreach $thing (@linearray)
    {
    $wordlist{$thing} = "";
    }
  }


while ((my $key, my $value) = each(%wordlist))
   {
   $key =~ s/\n//o;
   if ($key =~ /[a-z]+/)
      {
      print OUT "$key\n";
      }
   }
end;
```

LIST 2.6.2. ALPHABETIZED SAMPLE FROM `meshword.txt` **FILE**

```
lepeophtheirus
leper
lepidium
lepidoptera
lepidopterous
lepidozamia
lepilemur
lepilemurs
leponex
lepori
```

```
leporidae
leporids
leporipoxvirus
leporipoxviruses
leprae
lepraemurium
lepromatous
lepromin
leprosies
leprostatic
leprostatics
leprosum
leprosy
leprous
leptazole
leptin
leptoconops
leptocurares
leptolyngbya
leptomeningeal
leptomeninges
leptomeningitis
leptomonas
leptonema
leptophos
```

Word lists are simple to create and have many different uses (List 2.6.3).

LIST 2.6.3. USES FOR PERL ROUTINES THAT CREATE WORD LISTS FROM DOCUMENTS

Create spell checkers.
Find misspelled words in a text.
Implement approximate (also called fuzzy) matching algorithms.
Develop de-identification protocols.
Write word-substitution scripts.
Quantify the word-complexity of a text.

CHAPTER

3

File/Directory Operations and Text Transformations with Regular Expressions

3.1. BACKGROUND

For many biomedical professionals, computational tasks will be limited to transforming textual records into a preferred format or a preferred method of organization (List 3.1.1)

LIST 3.1.1. EXAMPLES OF FILE TRANSFORMATION NEEDS OF BIOMEDICAL PROFESSIONALS

1. The state's tumor board has issued a new format for reporting new cancer cases. Old cancer registry datasets must be reconfigured to the new format.
2. A memo from top executives indicates that all legacy text files should be prepared as html documents. Three months later, another memo requires all html documents to be reformatted as XML documents.
3. A directive from the IT department requires all dates to conform to the ISO 86015 formats for dates and times. Legacy records and other medical files must be rewritten with dates and times that conform to ISO 86015.
4. An editor requires authors to submit supplemental material supporting experimental findings described in manuscripts. The supplemental material must conform to a published data-exchange specification. Excel files will not be accepted.
5. In a moment of zeal, the institution's network security office asks you to provide a list of every `.exe` file on your computer.
6. From a series of relational data tables, you need to create a single, merged, hierarchical data file.

A common feature of the listed tasks is that none of them can be accomplished using a commercial software application. All tasks are entangled in the special needs of the user and do not yield to a general solution.

One of the acronyms often applied to Perl is Practical Extraction Report Language. More precisely, this is a retro-acronym (assigned after the name "Perl" was invented). Another, less flattering, acronym for Perl is Pathologically Eclectic Rubbish Language. History aside, Perl excels at collecting files from directories, parsing through collections of files, and transforming files into other files that have a desired revised content, organization, or format.

The purpose of this chapter is to show how Perl finds, opens, and parses files and how simple Perl commands can transform and reorganize files while they are being parsed.

3.2. LISTING FILES IN A DIRECTORY WITH `readdir`

Often, what you need to do to a file depends on what kind of file you have encountered. Perl has many file test operators that test file properties (List 3.2.1).

LIST 3.2.1. SOME PERL FILE TEST OPERATORS

```
-e  File exists
-s  File has nonzero size (returns size)
-f  File is a plain file
-d  File is a directory
-T  File is a text file
-B  File is a binary file (opposite of -T)
```

A simple `readdir` command accesses directory files. Using the `readdir` command and the file test operators, you can easily obtain information about the files in the current directory (List 3.2.2).

In DOS, the current directory is named "." by convention:

```
opendir (MYDIR, ".") || die ("Can't open directory");
```

This line opens the current directory for reading (by the `readdir` command). The meat of the script is in the block line:

```
print "$file\n" if (-f $file && -T $file);
```

LIST 3.2.2. PERL SCRIPT dirtest.pl, FOR TESTING FILES IN THE CURRENT DIRECTORY

```
#!/usr/local/bin/perl
opendir (MYDIR, ".") || die ("Can't open directory");
while ($file = readdir (MYDIR))
  {
  print "$file\n" if (-f $file && -T $file);
  }
closedir (MYFILE);
exit;
```

The block line tells Perl to test a file to see if it is a plain file and a text file. If it is, the name of the file is printed to a line of output on the monitor.

3.3. LISTING FILES IN A DIRECTORY AND ITS SUBDIRECTORIES WITH THE File::list MODULE

The limitation of the readdir command is that it finds files only in a specified directory. It has no iterative facility for searching branched subdirectories. This is where the File::List module comes in handy.

The script filelist.pl looks in the D directory and lists all the files from all the subdirectories, along with filesize and date that the file was last modified (List 3.3.1).

LIST 3.3.1. PERL SCRIPT filelist.pl, LISTS FILES FROM DIRECTORIES AND SUBDIRECTORIES

```
#!/usr/bin/perl
use File::List;
my $search = new File::List("d\:");
print "What output file would you like?";
$filename = <STDIN>;
$filename =~ s/\n//;
open (OUT, ">$filename");
$search->show_empty_dirs();
my @files  = @{ $search->find("\.+\$") };
$count = 0;
foreach $thing (@files)
```

```
    {
    $count++;
    @stat = stat($thing);
    $size = $stat[7];
    print "$thing $size\n";
    }
print "The number of files is $count\n";
exit;
```

The script prompts the user to supply the name of the output file to receive the list.

It also uses the `File::List` module, which can be installed easily using activestate's ppm downloading facility.

3.4. CONCATENATING TEXT FILES

This little script creates one file from a list of files (List 3.4.1).

This script introduces several Perl functions.

The `binmode()` function tells Perl to treat a file as though it were a binary file. When you are handling a file as a unit and have no intention of reading sequential lines from the file, using a `binmode` command is a wise precaution:

```
binmode (HOLD);
```

This script makes coordinated use of the `seek()`, `tell()`, and `read()` commands:

```
seek (LISTFILE, 0, 2);
$size = tell(LISTFILE);
seek(LISTFILE,0,0);
read(LISTFILE, $lump, $size);
seek(LISTFILE, 0, 2);
```

The syntax for the seek function is:

```
seek (FILEHANDLE, OFFSET, FROM);
```

When the FROM parameter is set to 0, it refers to the beginning of the file. When the FROM parameter is set to 1, it refers to the current position in the file.

When the FROM parameter is set to 2, it refers to the end of the file.

LIST 3.4.1. PERL SCRIPT `concat.pl`**, CONCATENATING TWO FILES**

```
#!/usr/local/bin/perl
open(HOLD,">holder.txt")||die"cannot";
binmode (HOLD);
@filelist = ("cdeout", "bwtpl\.txt", "bwt\.pl");
#list of files. Substitute your own files in @filelist.
foreach $file (@filelist)
  {
  open (LISTFILE, $file)||die"cannot";
  binmode (LISTFILE);
  seek (LISTFILE, 0, 2);
  $size = tell(LISTFILE);
  seek (LISTFILE,0,0);
  read(LISTFILE, $lump, $size);
  print HOLD $lump;
  close LISTFILE;
  }
close HOLD;
#we are done, but let's check that we can retrieve a file
@filestats = stat("cdeout");
$size = $filestats[7];
open (HOLD, "holder.txt");
binmode (HOLD);
read(HOLD, $firstfile, $size);
print $firstfile;
end;
```

The offset is the number of bytes to move from the FROM position.

To indicate that the file position is set 100 bytes before the end of the file use the following command:

```
seek(HOLD, -100, 2);
```

To indicate that the file position is set ahead 30 bytes from the current position, use the following command: `seek(HOLD, 30, 1);`

We used the seek command to set the position to the end of the file:

```
seek (LISTFILE, 0, 2);#sets read position to end of file
$size = tell(LISTFILE);#provides the current file position
seek (LISTFILE,0,0); #sets read position to beginning of file
read(LISTFILE, $lump, $size);
```

The `tell()` command gives us the current position in the file. Following a `seek()` command sending the position to the end of the file, the `tell()` command returns the size of the file.

The `read()` command reads the number of bytes specified by the last variable (in this case, *$size*), from the designated filehandle (in this case `LISTFILE`), starting from the current position in the file (set at the beginning of the file in the prior command), and puts the resultant string into the middle variable (in this case *$lump*).

The remainder of the script is straightforward. One new trick is provided, however:

```
@filestats = stat("cdeout");
$size = $filestats[7];
```

Perl's `stat()` function returns an array of information about a specified file. We do not need to discuss all the information returned by the `stat()` function. Suffice it to say that the seventh array element is the size of the file. Some programmers find this method for determining file size more convenient than a `seek()` followed by a `tell()`.

3.5. MORE PERL MODULES FOR FILE OPERATIONS

We have used several Perl modules specialized for file operations. A list of some of the more important file modules is shown (List 3.5.1).

LIST 3.5.1. SOME PERL MODULES FOR FILE OPERATIONS AND TRANSFORMATIONS

```
File::compare
File::copy
File::find
File::sort
Spreadsheet-parseexcel
XML-excel
RTF-document
```

3.6. EXTRACTING LINES OF TEXT THAT MATCH A REGULAR EXPRESSION

Many Perl scripts involve opening a file, parsing each line of text, and selecting those lines that match a pattern of interest. This script parses every line of a file, looking for the phrase "Dr. notified" (List 3.6.1).

LIST 3.6.1. PERL SCRIPT `extract.pl` SELECTS LINES FROM A FILE THAT MATCH A PATTERN

```
#!/usr/bin/perl
open (TEXT, "filename.txt"); #use a preferred filename
open (OUT, ">list\.out");
$line = " ";
while ($line ne "")
  {
  $line = <TEXT>;
  if ($line =~ /Dr\. notified/)
    {
    print OUT $line;
    }
  }
exit;
```

Scripts, such as `extract.pl`, have many uses in biomedicine. There are frequent occasions when you need to pull the files, records, or reports that pertain to a particular subject or that contain a particular phrase (for example, "No evidence of malignancy") or a particular value (for example, glucose above 110). A parsing script with a regular expression "if block" is often the solution.

In the `extract.pl` script, we could have jiggled the regex to include the word "Dr." followed by any other text, followed by the work "notified" to allow the inclusion of the doctor's name in the matching text.

```
if ($line =~ /Dr\.?.+notified/i)
```

In this example, the first dot is meant to represent the period following "Dr". A slash precedes the dot to signify that the dot represents a character, not a Perl operator. The "?" following the dot tells Perl that the dot may or may not occur. This is followed by ".+". Here, we use the dot as a `regex` operator indicating "any character" and the "+" tells us that "any character" can occur one or more times. Following the pattern, we add the "i" option, indicating that the match can be case-insensitive.

We could actually shorten the regular expression:

```
if ($line =~ /Dr.+notified/i)
```

Because we do not care whether a period follows the "Dr", we can omit the requirement that a period may or may not appear in the matching string.

In Chapter 6.3, we will use the script `proxfind.pl` to show how proximity searches will look for patterns that fall within a specified number of characters from each other.

3.7. ASCII CHARACTERS

There are 256 (2^8) different characters in the ASCII character set. ASCII values under 128 are often referred to as 7-bit ASCII. All of the printable characters (that is, the characters that a keyboard can print) are assigned an ASCII number under 128 and can all be specified in 7 bits (List 3.7.1).

LIST 3.7.1. SEVEN- AND EIGHT-BIT ASCII CHARACTERS

```
128 in binary is  2^7        10000000
127 in binary is              1111111
256 in binary is  2^8 or 100000000
255 in binary is             11111111
```

All of the printed characters found in this book are ASCII characters from 33 to 126 (List 3.7.2). The unprinted characters in this book are the space and the doublet carriage-return/line feed. These three unprinted characters are ASCII 32 (space), ASCII 13 (carriage return), and ASCII 10 (line feed).

LIST 3.7.2. THE ALPHANUMERIC SEVEN-BIT ASCII CHARACTERS

32	space	49	1
33	!	50	2
34	“	51	3
35	#	52	4
36	$	53	5
37	%	54	6
38	&	55	7
39	‘	56	8
40	(	57	9
41	)	58	:
42	*	59	;
43	+	60	<
44	,	61	=
45	-	62	>
46	.	63	?
47	/	64	@
48	0	65	A

66	B	97	a
67	C	98	b
68	D	99	c
69	E	100	d
70	F	101	e
71	G	102	f
72	H	103	g
73	I	104	h
74	J	105	i
75	K	106	j
76	L	107	k
77	M	108	l
78	N	109	m
79	O	110	n
80	P	111	o
81	Q	112	p
82	R	113	q
83	S	114	r
84	T	115	s
85	U	116	t
86	V	117	u
87	W	118	v
88	X	119	w
89	Y	120	x
90	Z	121	y
91	[	122	z
92	\	123	{
93	]	124	|
94	^	125	}
95	_	126	~
96	`		

3.8. STRIPPING UNPRINTABLE ASCII CHARACTERS USING `tr`

Occasionally, you might find it useful to eliminate characters from files that do not correspond to standard print characters. There are many ways to strip such characters, but the `tr` operator offers a slick and fast solution (List 3.8.1).

LIST 3.8.1. PERL'S `tr` OPERATOR TRANSLATES ANY CHARACTER OR RANGE OF CHARACTERS TO ANY OTHER CHARACTER OR RANGE OF CHARACTERS

```
$string =~ tr/A/a/;      #Translates occurrences of A to a
$string =~ tr/q/z/;      #Translates occurrences of q to z
$string =~ tr/A-Z/a-z/; #Translates uppercase letters to
                         #lowercase
$string =~ tr/0-9/ /c;  #Uses the complement option to
                         #translate everything that is not
                         #a digit to a space
$string =~ tr/0-9//d;   #Uses the delete option to delete digits
$string =~ tr/0-9//cd;  #Uses both the delete and the
                         #complement option to
                         #delete everything that is not a digit
```

The slash before a number is interpreted as an octal form, and the `tr` operator automatically interprets an octal number as an ASCII equivalent. A \040 is (8*4) or 32, and a \176 is (1*64) + (7*8) + 6 = 126. The following command deletes all ASCII characters in the range of 128 through 255, thus eliminating all 8-bit ASCII characters from the string:

```
$string =~ tr/\040-\176//cd;
```

A minor modification will preserve the newline character:

```
$filecontent =~ tr/\12\15\040-\176//cd;
\12 is the octal equivalent of 10 or the line-feed character
\15 is the octal equivalent of 13 or the carriage-return character
```

LIST 3.8.2. PERL SCRIPT `keys.pl`, STRIPS ALL NON-KEYBOARD CHARACTERS FROM A FILE

```
#!/usr/local/bin/perl
open (TEST, "mums.jpg");
undef($/);
$filecontent = <TEST>;
$filecontent =~ tr/\040-\175//cd;
print $filecontent;
exit;
```

The line-feed character and the carriage-return character combine as the newline character, which is preserved, along with the characters in the ASCII range 32-126 in the modified `tr` operator.

3.9. CONVERTING BINARY TO BASE64

Though we distinguish text files from binary files, all files are actually binary files. Sequential bytes of eight bits are converted to ASCII equivalents, and if the ASCII equivalents are alphanumerics, we call the file a text file. If the ASCII values of eight-bit sequential file chunks are non-alphanumeric, we call the files binary files.

Any file can be converted into something akin to a text file by dividing the file into six-bit chunks, and assigning each six-bit chunk an alphanumeric ASCII character with two leading zeros (that is, a 0-padded six-bit ASCII value, which is equivalent to a Base64 character.). Binary can be interconverted with Base64. Base64 conversion is sometimes used to represent a binary file as a Base64 alphnumeric file. Alphanumeric files can be ported inside formats that require plain ASCII text (such as html and xml files).

A short script for converting binary jpeg files to Base64, `jpg2b64.pl` is shown later in List 18.13.1.

3.10. JUSTIFYING A TEXT FILE

In printing, justification is the process whereby spacing within a line is adjusted so that each line on a page ends evenly, producing a straight right-side margin.

Neatness aside, text justification can be important when an ASCII file has no line breaks (for example, the one-line file), lines of vastly different lengths, or lines that cannot fit on a typed page or a monitor, or when ASCII files of different line lengths are merged. Whenever you read an e-mail that has lines that extend beyond the monitor screen or lines abruptly chopped in the middle of the page, you are the innocent victim of an unjustified textual assault.

In ASCII, spaces are all of fixed length. There is no mechanism to widen the length of a space, other than padding additional spaces. Adding spaces between some words, and not others, to justify a line of text, is esthetically unacceptable. However, it is simple to write a Perl script that will do a fair job at justification.

This handy script parses through text files, combining all the lines of a paragraph into a single line and then chopping the paragraph into neat lines that never exceed 80 characters in length and that never break words (List 3.10.1). The script can be easily modified to produce any chosen line length.

LIST 3.10.1. PERL SCRIPT `paraform.pl` ADDS CARRIAGE RETURNS TO THE LINES OF A PARAGRAPH

```
#!/usr/local/bin/perl
print "What file would you like to reformat?\n";
$filename = <stdin>;
chomp($filename);
open (TEXT, $filename);
$filename =~ s/.$/\_/o;
open (NEWTEXT, ">$filename");
undef($/);
$line = <TEXT>;
$line =~ s/\n\n+/ \*\-\*\-\*/g;
$line =~ s/\n/ /g;
@textarray = split(/\*\-\*\-\*/,$line);
foreach $thing (@textarray)
   {
   @eighty_line=grep{length()>0}split(/(.{1,80}) /,$thing);
   foreach my $value (@eighty_line)
            {
            $value =~ s/^ +//o;
            print NEWTEXT  "$value\n";
            }
    print NEWTEXT "\n\n";
    }
exit;
```

Early in the script, the input-record separator is undefined, and the entire file is slurped into a variable (*`$line`*):

```
undef($/);
$line = <TEXT>;
```

We accept runs of more than one newline character to indicate the beginning of a paragraph or of a line that is intentionally set apart from other lines in the text. Wherever we see two or more newline characters together, we replace these with a space followed by a pattern that we expect is unique and does not occur elsewhere in the text (`*-*-*`):

```
$line =~ s/\n\n+/ \*\-\*\-\*/g;
```

The singly occurring newline characters can be eliminated and replaced with a space:

```
$line =~ s/\n/ /g;
```

We can then split the text at every occurrence of `*-*-*`, putting the paragraphs into an array:

```
@textarray = split(/\*\-\*\-\*/,$line);
```

Most of the work of the paraform.pl script is accomplished in one line:

```
@eighty_line=grep{length()>0}split(/(.{1,80}) /,$thing);
```

Each paragraph can be split using the /(.{1,80}) / pattern. Remember that the unescaped "." appears in a Regex, which matches any character other than the newline character. Perl pattern matches are, by default, greedy. Perl will look for the longest stretch of string characters that will satisfy the pattern. In this case, Perl will look for the longest stretch of characters, up to 80, that are followed by a space, splitting array items on the matched pattern. This divides a succession of words into neat lines never exceeding 80 characters and never chopping an end word. The `grep` operator tests the array of text lines to ensure that no lines are empty, and then passes the text lines to the `@eighty_line` array.

3.11. PREPARING ASCII FILES FOR IMPORT TO WORD PROCESSORS

ASCII files seldom import neatly into word processors because they have too many return characters between lines. Word processors almost all have word-wrap, so that the lines are seamlessly wrapped into a nice-looking paragraph without carriage returns between lines. To convert a file with hard carriage returns into one without carriage returns (so that it can be slurped into a good-looking word processor file), you may use a modified `towpd.pl` (List 3.11.1). The only carriage returns that are saved are the double-returns between paragraphs.

3.12. FINDING ALL THE MEDICAL ABBREVIATIONS IN A FILE

Most medical abbreviations were no doubt invented as time-savers for health professionals. These same abbreviations can waste the time of persons tasked with organizing, indexing, searching, or interpreting medical text. Such persons reading a medical record are expected to know that "ga" occurring within an obstetric history usually means "gestational age," while "ga" occurring within a dermatologic history usually means "granuloma annulare." Automatic indexers and machine

LIST 3.11.1. PERL SCRIPT `towpd.pl` REMOVES HARD RETURNS FROM LINES IN PARAGRAPHS

```
#!/usr/local/bin/perl
print "What file would you like to translate?";
$filename = <STDIN>;
chomp($filename);
open (TEXT, $filename)|| die "No file found!";
open (HOLDFILE, ">holder.txt");
$hold = " ";
while ($hold ne "")
  {
  $hold = <TEXT>;
  if ($hold eq "\n")
     {
     print HOLDFILE "\n\n";
     next;
     }
  $hold =~ s/\n/ /g;
  print HOLDFILE $hold;
  }
close TEXT;
close HOLDFILE;
exit;
```

translators of medical text require large and accurate lists of medical abbreviations and accurate algorithms to find abbreviations within text and to map the abbreviations to their correct expansions.

In this section, we will describe a Perl script that finds abbreviations in a text corpus and annotates them with all known expansions of the abbreviation.

We'll be using the author's public domain abbreviation file, `berm_abb.txt` (2) a sample of which is shown in List 3.12.1.

The surprising output of `abbadd.pl` (List 3.12.2) serves as a stern lesson for biomedical professionals (List 3.12.3). A single abbreviation may have many different expansions. In the example text, "pa" was probably intended to mean "posteroanterior." Those who read medical charts know that occasions arise when experts cannot disambiguate an

LIST 3.12.1. A SAMPLING OF THE `berm_abb.txt` **LIST OF ABOUT 12,000 MEDICAL ABBREVIATIONS**

```
arm = artificial rupture of membrane (caps required)
   (additional membranes)
armd = age related macular degeneration
arn = acute retinal necrosis
aroa = autosomal recessive ocular albinism
arom = artificial rupture of membranes (alternate membrane)
arp = american registry of pathology
arpkd = autosomal recessive polycystic kidney disease
arsa = arylsulfatase a
art = anti-retroviral therapy (caps required)
art = arterial (alternate artery) (caps required)
arter = arterial
aru = acute retention of urine
arv = aids-related virus
arv = anti-retroviral therapy
arvc = arrhythmogenic right ventricular cardiomyopathy
arvd = arrhythmogenic right ventricular dysplasia
   (additional familial)
as = angelman syndrome (caps required)
as = aortic stenosis (caps required)
as = arteriosclerosis (alternate arteriosclerotic) (caps
   required)
as = auris sinestra (alternate left ear) (caps required)
asa = acetylsalicylic acid
asa = argininosuccinic acid
```

LIST 3.12.2. PERL SCRIPT `abbadd.pl` **ANNOTATES PHRASES WITH THEIR ABBREVIATIONS**

```
#!/usr/local/bin/perl
open (HASHER, "berm_abb.txt")||die"Can't open file";
open (OUT, ">output.txt")||die"Can't open file";
$hashline = " ";
while ($hashline ne "")
  {
  $hashline = <HASHER>;
```

```
    $line = $hashline;
    next if ($line !~ /\=/);
    $line =~ s/\n//;
    $line =~ s/\([^\)]+\)//g;
    $line =~ / \= /o;
    $abb = $`;
    $expand = $';
    $expand =~ s/^ +//;
    $expand =~ s/ +/ /g;
    $expand =~ s/ +$//;
    if (exists($item{$abb})) #checks to see if the hash key exists
      {
      $item{$abb} = $item{$abb} . "\, $expand";
      }
    else
      {
      $item{$abb} = $expand;
      }
    }
close HASHER;
$text = "The pa view of the cxr indicated that the bx would
   involve the lul. The 38 yo wm patient was still waiting in
   the er for the results of his cxr.";
$textcopy = $text;
$textcopy = lc($textcopy); #makes everything lowercase
@textwords = split(/ +/,$textcopy);
@textwords = sort (@textwords);
$oldword = "";
foreach $word (@textwords)
   {
   next if ($word eq $oldword);
   if (exists $item{$word})
     {
     $text =~ s/\b$word\b/$& \($item{$word}\)/ig;
     }
   $oldword = $word;
   }
print OUT $text;
exit;
```

LIST 3.12.3. THE SURPRISING OUTPUT OF `abbadd.pl`

```
The (tetrahydrocortisone) pa (panic attack, pantothenic
    acid, paranoia, pascal, pathologist, pathology, penetrated
    posteroanterior, pennsylvania, perianal, periapical,
    periodic acid, peripheral arteriosclerosis, pernicious
    anemia, phenylalanine, phenylalkylamine, phosphatidic acid,
    phosphoarginine, photoallergy, phthalic anhydride, physician
    assistant, plasminogen activator,
    platelet aggregation, pleomorphic adenoma, polyamine,
    posteroanterior, prealbumin, primary aldosteronism, primary
    amenorrhea, proanthocyanidin, procainamide, program
    announcement, propionic acid, protactinium, protrusio
    acetabuli, psoriatic arthritis, psychoanalyst, pulmonary
    artery, pulmonary atresia, puromycin aminonucleoside,
    pyrrolizidine alkaloid, pyruvic acid) view of the
    (tetrahydrocortisone) cxr (chest xray) indicated that the
    (tetrahydrocortisone) bx (biopsy) would involve the
    (tetrahydrocortisone) lul. The (tetrahydrocortisone) 38 yo
    (year old) wm (white male) patient was (wiskott aldrich
    syndrome) still waiting in (inches, index, indiana, indium,
    insulin, interstitial nephritis, inulin) the
    (tetrahydrocortisone) er (electroresection, emergency room,
    endoplasmic reticulum, epigastric region, erythrocyte,
    erythrocyte receptor, esophageal rupture, estradiol
    receptor, estrogen receptor, estrogen receptors, external
    rotation) for the (tetrahydrocortisone) results of his
    (hospital information system) cxr (chest xray).
```

abbreviation with multiple expansions (List 3.12.4). For instance, can "MVR" be confidently expanded to "mitral valve repair" when it can also mean "mitral valve replacement?" Can "IBD" be confidently expanded to "inflammatory bowel disease" when it can also mean "idiopathic bowel disease?" How many people have been prematurely dropped from hospital and federal databases after a charted "DOD" was interpreted as "Date of Death" when its intended meaning was "Date of Discharge?"

The output of `abbadd.pl` also shows us that common terms, such as "the," "in," and "was" can also serve as medical abbreviations (List 3.12.5).

LIST 3.12.4. EXPANSIONS OF "PA"

panic attack, pantothenic acid, paranoia, pascal, pathologist, pathology, penetrated posteroanterior, pennsylvania, perianal, periapical, periodic acid, peripheral arteriosclerosis, pernicious anemia, phenylalanine, phenylalkylamine, phosphatidic acid, phosphoarginine, photoallergy, phthalic anhydride, physician assistant, plasminogen activator, platelet aggregation, pleomorphic adenoma, polyamine, posteroanterior, prealbumin, primary aldosteronism, primary amenorrhea, proanthocyanidin, procainamide, program announcement, propionic acid, protactinium, protrusio acetabuli, psoriatic arthritis, psychoanalyst, pulmonary artery, pulmonary atresia, puromycin aminonucleoside, pyrrolizidine alkaloid, pyruvic acid

It is tempting to think that an abbreviation can be distinguished from a word homograph based on its "sense" in the sentence or by some consistent format (for example, uppercase letters or periods between letters). My experience has been that medical narrative text is unpredictable, inconsistent, ungrammatical, and nearly senseless. The task of deriving meaning from abbreviations embedded in medical text remains a challenge.

Let us review the `abbadd.pl` script (List 3.12.1). First, Perl passes through all lines in the `berm_abb.txt` file containing abbreviations in the form:

asa = argininosuccinic acid

Perl matches for an "=" character flanked on both sides by a single space:

```
$line =~ / \= /o;
```

When Perl performs a match, it automatically creates two special variables that the programmer may find handy:

$`, the so-called prematch, consisting of the string characters that occur prior to the matching pattern

$', the so-called postmatch, consisting of the string characters that follow the matching pattern

LIST 3.12.5. EXPANSIONS OF "IN"

inches, index, indiana, indium, insulin, interstitial nephritis, inulin

In this case, the abbreviation itself is the prematch string and the expansion is the postmatch string. New variable names are assigned the prematch and postmatch strings:

```
my $abb = $`;
my $expand = $';
```

Remember, we intended to build a hash consisting of abbreviations as the hash keys and a list of expansions as the hash values. We name the hash %item and build new key/value pairs:

```
if (exists($item{$abb}))
  #checks to see if the hash key exists
  {
  $item{$abb} = $item{$abb} . "\, $expand";
  }
else
  {
  $item{$abb} = $expand;
  }
```

As the script loops through each line of the abbreviations file, it will add multiple values to single hash keys.

For instance, there are two expansions for the abbreviation fb:

fb = fluid balance

fb = foreign body

When the script reaches the first entry, it tests to see if the fb key already exists. Because it does not, it creates the hash key/value pair:

```
$item{$fb} = "fluid balance"
```

On the next loop, it encounters the key fb again. This time, the `exists()` condition tests as true. In this case, the key/value pair is:

```
$item{$abb} = $item{$abb} . "\, $expand";
```

which evaluates to:

```
$item{"fb"} = $item{"fb"} . ", foreign body"
```

or, after concatenation with the "." operator,

```
$item{"fb"} = "fluid balance, foreign body";
```

In this manner, we build a hash that collects all of the expansion for an abbreviation in the same key/value pair. Once the hash of abbreviations is created, the next step is to collect all of the words in the sample text and to add the list of abbreviation expansions to the text.

```
 foreach $word (@textwords)
 {
next if ($word eq $oldword); #we will skip duplicates
 if (exists $item{$word})
   {
   $text =~ s/\b$word\b/$& \($item{$word}\)/ig;
  #$& is Perl's special variable for the matched text
   }
$oldword = $word; -
#$oldword tells the next iteration if
#the word has already been looped
   }
```

The array `@textwords` contains the sorted list of words in the text. The sorted list of words might contain duplicates, so we check at the top of each loop to ensure that the loop is examining a new word.

If the looped word (*$word*) exists as a key in the abbreviation hash (`%item`), we will go back to the original text and substitute the word in the text for the word and its list of expansions.

We're finished! The program has created an output that lists all of the possible expansions for text words that match entries in our master list of abbreviations.

3.13. FILTERING NEGATIONS

It is sometimes important to distinguish a positive assertion from a negative assertion. This is particularly true in medicine, where tests are often ordered to rule out diseases. For instance, if a patient presents with a persistent fever, a physician might order tests for a collection of diseases known to cause fevers of unkown origin. Negative test findings often rule out most of the suspect diseases, permitting the physician to narrow her attention to one or several entities.

Consequently, medical text is replete with negation phrases (List 3.13.1).

LIST 3.13.1. EXAMPLES OF NEGATION PHRASES COMMONLY FOUND IN MEDICAL REPORTS

Tumor not found.
Absence of lymphadenopathy or splenomegaly.
Margins free of basal cell carcinoma.
No yeast or hyphae seen.

All of these terms are negative assertions, but not all of the terms contain a negating adjective or noun, such as "not" or "negative." For instance, "Margins free of basal cell carcinoma," seems like a positive assertion unless you understand that the statement negates the presence of basal cell carcinoma at the margin.

Computer programs that parse text and count the occurrence of diagnostic entities reported within the text, must distinguish positive assertions (for example, tumor is present) from negative assertions (for example, tumor is not present). Otherwise, the incidence statistics assembled by the parsing program will be erroneous (3).

A simple negation parser follows (List 3.13.2).

LIST 3.13.2. PERL SCRIPT `neg_out.pl` **REMOVES NEGATIONS FROM TEXT**

```
#!/usr/bin/local/perl
use Encode;
use Sentence;
open (OUT, ">negtext.txt");
$inputfilename = "med_text.txt";
$trialfile = Sentence->new();
$outputfilename = $trialfile->sentence_parse($inputfilename);
open (TEXT, $outputfilename)||die"Cannot open file";
my $line = " ";
while ($line ne "")
  {
  $count++;
  $line = <TEXT>;
  $line = lc($line);
  $line =~ s/\.\n/\n/o;
my @negativearray = qw(neither nor inconsistent  fails no
   failed insufficient incompatible  inconsistent not absent
   absence negative);
  foreach $value (@negativearray)
     {
     if ($line =~ /\b$value\b/i)
        {
        print OUT "$count\-$line";
        }
     }
  }
exit;
```

Early in the script, an external module, Sentence, is called upon to break an input file into discrete sentences:

```
      $inputfilename = "med_text.txt";
$trialfile = Sentence->new();
$outputfilename = $trialfile->sentence_parse($inputfilename);
```

We will return to the Sentence module much later, in Chapter 17. For now, suffice it to say that Sentence is a class module that contains instructions for its proper use. In these three lines, a file is passed to the class module, and another file is returned, with each sentence from the original file occupying a separate line in the output file. This allows us to parse the text sentence-by-sentence. If we find a negation in the sentence, we simply will not count any disease terms that are included in the negated sentence.

We place all of our negation modifiers into an array:

```
     my @negativearray = qw(neither nor inconsistent fails no
n't failed insufficient incompatible inconsistent not absent
absence negative);
```

The remainder of the script is straightforward:

```
. foreach $value (@negativearray)
    {
    if ($line =~ /\b$value\b/i)
        {
        print OUT "$count\-$line";
        }
    }
}
```

The script parses through the lines (now sentences) of the text looking for matches against any of the negation terms. If there is a match, the script extracts the line. A human can review the extracted lines to determine whether the extracted lines indeed contain negation statements and whether there might be positive statements included among the negations.

The kinds of negated sentences extracted by the script would include the listed negation terms (List 3.13.3).

This script is far from perfect, but it demonstrates an approach to text parsing that can be developed with a few lines of Perl.

3.14. EXTRACTING AN E-MAIL LIST FROM A PUBMED SEARCH

PubMed is the National Library of Medicine's public database of medical journal articles. PubMed is an expansion of Medline and contains

LIST 3.13.3. SOME MATCHED NEGATION TERMS

is absent	absence of
is not found	is ruled out
is not seen	we can rule out
is not present	not in evidence
no evidence of	not consistent with
no presence of	inconsistent with
can not find	incompatible with
cannot find	insufficient evidence
neither	failed to show
nor	fails to
without	are negative
negative for	no features of
is excluded	zero evidence for
inconclusive	

journal abstracts and a set of sophisticated search routines. PubMed also links open-access journal articles to Web sites from which the complete text can be downloaded.

Every biomedical professional should be adept at using PubMed. PubMed provides Perl programmers with a great many opportunities to acquire, search, transform, and organize large text files of abstracts and other PubMed citation data.

As an exercise, go to the PubMed Web site, and enter a search on the word "informatics" (see Figure 3-1). Click on the "GO" button, and the first 20 citations (from over 7500 citations) will appear on your screen. PubMed lets you change the display format for citations. From the "Display" button, pick "MEDLINE". Your screen will refresh with citations listed in longer entries that include each journal article's abstract and a variety of annotative data in a uniform format. Next, click on the "Send To" button and chose "File". PubMed will download to your hard drive a file containing 7500+ citations in MEDLINE format. The file size for this particular search should be in excess of 13 mbytes.

The resultant citation file, in Medline format, is a perfect crucible for data-mining projects. For a start, you might want to extract the names, street addresses, and e-mail addresses for all of the authors in the citation files for whom e-mails are listed (not all citations will contain e-mail addresses) (List 3.14.1).

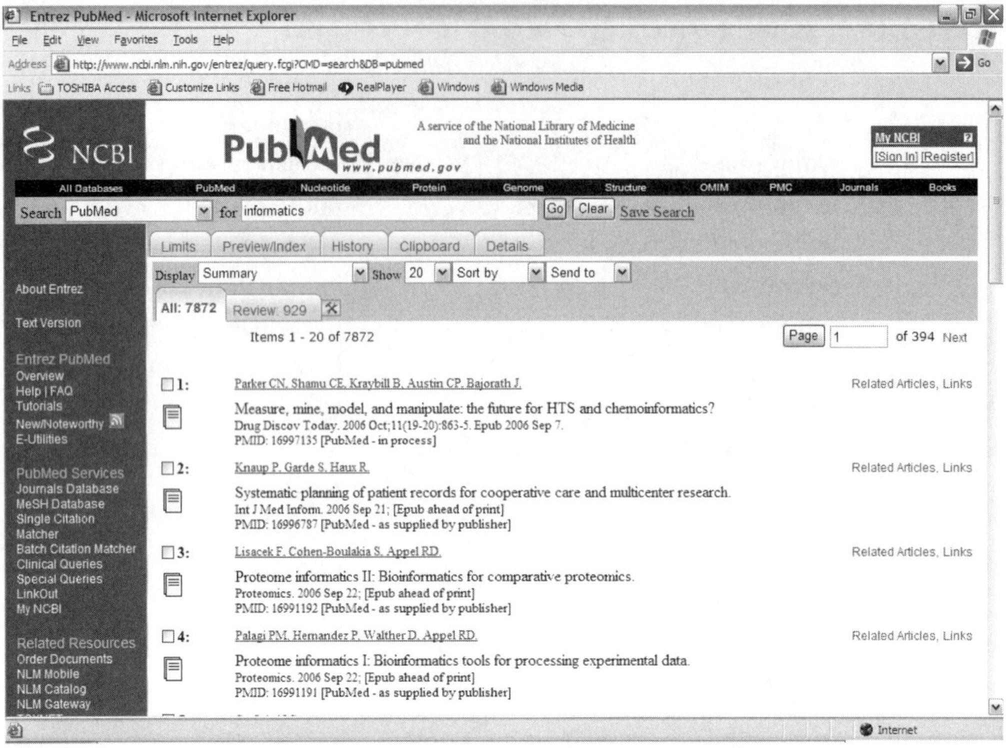

Figure 3-1 PubMed search results.

LIST 3.14.1. PERL SCRIPT `epullus2.pl` EXTRACTS E-MAIL ADDRESSES AND STREET ADDRESSES FROM A PUBMED SEARCH.

```
#!/usr/bin/perl
open(OLD,"sent.txt")||print"No exclusion file";
open(TEMP, ">e_names.txt")||die"Can't";
open(NEW, ">email.txt")||die"Can't";
$oldnames = " ";
while ($oldnames ne "")
    {
    $oldnames = <OLD>;
    print TEMP $oldnames;
    print $oldnames;
    $oldnames =~ s/\n//;
    $oldhash{$oldnames} = 1
    }
print "What file do you want to read?";
```

```
$filename = <STDIN>;
chomp($filename);
open (TEXT, $filename)||die"Can't open file";
$/ = "PMID-";
$line = " ";
while ($line ne "")
    {
    $line = <TEXT>;
    if ($line =~ /FAU \- ([\w\'\.\, ]+)\n/mo)
     {
     $name = $1;
     }
    $line = lc($line);
    if ($line =~ /\b[\w\d\_\-\.\~]+\@[\w\d\_\-\.\~]+\b/)
     {
     $line = $&;
     next if (exists $oldhash{$line});
     $oldhash{$line} = " ";
     $line = $name . " \| " . $line;
     $newhash{$line}++;
     #$newhash{$line}++ if ($line =~ /com/);
     #$newhash{$line}++ if ($line =~ /edu/);
     #$newhash{$line}++ if ($line =~ /gov/);
     #$newhash{$line}++ if ($line =~ /net/);
     #$newhash{$line}++ if ($line =~ /org/);
     }
    else
     {
     next;
     }
    }


while(($key, $value) = each (%newhash))
  {
  print "$key\n";
  print TEMP "$key\n";
  $key =~ /\| (.+)$/;
  $key = $1;
  print NEW  "$key\n";
  }
exit;
```

This script is particularly useful for creating e-mail lists for a possible group mailing. It begins by finding the file "sent.txt", which is an optional file listing e-mail addresses that you might want to exclude from a mailing. The sent.txt file might contain addresses that you have previously included in a group mailing. To avoid earning a reputation as a spammer, you might want to exclude names from a prior mailing in the current mailing. The epullus2.pl script looks for the sent.txt file. If there is a sent.txt file, it prepares a hash of the e-mail addresses in the file. Otherwise, it simply prints to the monitor, "No exclusion file" and proceeds through the script.

The heart of the file is:

```
        if ($line =~ /\b[\w\d\_\-\.\~]+\@[\w\d\_\-
\.\~]+\b/)
         {
         $line = $&;
         next if (exists $oldhash{$line});
         $oldhash{$line} = " ";
         $line = $name . " \| " . $line;
         $newhash{$line}++;
         }
```

The script checks the parsed line to see if it contains an e-mail pattern:

```
if ($line =~ /\b[\w\d\_\-\.\~]+\@[\w\d\_\-\.\~]+\b/)
```

It skips e-mail addresses that match any of the e-mail addresses contained in the sent.txt "exclusion" file:

```
next if (exists $oldhash{$line});
```

It prepares a string variable consisting of the name and the e-mail address demarcated by a vertical bar:

```
$line = $name . " \| " . $line;
```

It adds the new address to the "exclusion" hash, in case it is contained in another PubMed citation elsewhere in the parsed file:

```
$oldhash{$line} = " ";
```

It adds the e-mail address to %newhash, which will be used later in the script to yield the file of e-mail addresses:

```
$newhash{$line}++;
```

When preparing e-mail lists, it is sometimes preferable to collect just those e-mail addresses that belong to a certain domain.

The following commands can be used to add e-mail addresses that have a `.com`, `.edu`, `.gov`, `.net`, or `.org` domain:

```
$newhash{$line}++ if ($line =~ /\.com/);
$newhash{$line}++ if ($line =~ /\.edu/);
$newhash{$line}++ if ($line =~ /\.gov/);
$newhash{$line}++ if ($line =~ /\.net/);
$newhash{$line}++ if ($line =~ /\.org/);
```

The same approach can be used to extract web addresses, institutions, author names, titles, or any component of consistently formatted textual files, such as PubMed query results.

CHAPTER 4

Indexing Text

4.1. BACKGROUND

Students define an index as "the pages at the end of a textbook that won't be on the test." The main purpose of an index is to help people find passages of interest within the text. I have had conversations with intelligent people who have concluded that in the era of computers, the index has become obsolete. They know that anyone can simply do a "find" on a word or phrase, and the computer will instantly bring you to its location(s) in the document file.

When a computer user does an exact "find" on a phrase, she is missing closely related terms that share a common root. A search on "colon" will miss occurrences of "colonic" and "colons." A "wild card" search on colon* (colon followed by zero or more alternate alphabetic characters) will find "colonic" and "colons," but will also find "colonel" and "colonial" and will miss colitis. A "find" on "colon carcinoma" will miss occurrences of terms such as "colon adenocarcinoma," "adenocarcinoma of colon," "colonic carcinoma," "colonic adenocarcinoma," and "carcinoma of the colon." The "find" operation, standard on all modern word processors and document viewers, is unlikely to replace the venerable art of indexing.

Actually, indexing is much more important today than it ever was in the past. This is partly because computers allow us to create indexes that are more helpful than traditional book indexes. More importantly, modern indexes can organize and retrieve data from multiple sources at different levels of concept granularity. The sophisticated indexer can provide tools to minimize the chance that important concepts are overlooked, and maximize the chance that related concepts from different data sources can be connected to provide relational knowledge absent from the primary data sources.

In later chapters, the reader will learn how XML metadata expands and modifies our concept of indexing and links together textual data sources (like electronic medical records) with highly structured datasets (for example, biomedical databases). As expected, there are Perl modules available for all these biomedical activities.

The purpose of this chapter is to show how easy it is to create simple indexes using Perl.

4.2. CONCORDANCE

A concordance is a complete listing of all the words in a text, along with the locations in the text where they occur. Concordances are usually created for books of great significance, in which every word is deemed important. It is no surprise that the bible has been the subject of numerous efforts of scribes and scholars to create a concordance for the many versions and translations of the new and old testaments. It is difficult to imagine the enormity of creating a concordance "by hand." Using Perl, a concordance can be constructed for any book in about a second, with fewer than a dozen Perl command lines (List 4.2.1).

LIST 4.2.1. PERL SCRIPT `concord.pl` **CREATES A CONCORDANCE**

```
#!/usr/local/bin/perl
open (TEXT, "1DFRE10.TXT");
open (OUT, ">1DFRE10.OUT");
$line = " ";
while ($line ne "")
    {
    $cumline = "";
    for($i=0;$i<100;$i++) #assumes 100 lines per page
        {
        $line = <TEXT>;
        $cumline = $cumline . $line;
        }
    $page++;
    $cumline = lc($cumline);
    $cumline =~ s/[^a-z\-\']/ /g;
    @wordarray =  sort(split(/[\n\s]+/,$cumline));
@concordance = grep { $marked{$_}++; $marked{$_} == 1; }
    @wordarray;
    undef %marked;
```

```
    foreach $thing (@concordance)
        {
        $wordpage{$thing} = $wordpage{$thing} . " $page";
        }
    }
#The concordance is finished.
#The next lines just display it on-screen.

foreach $key (sort keys %wordpage)
    {
    print OUT "$key \= $wordpage{$key}\n";
    }
exit;
```

4.3. INDEXES

Indexes are somewhat different from concordances. An index is composed of words and phrases from a book that hold particular interest to readers, followed by their locations in the book. An indexing script is nearly identical to a script that creates a concordance (List 4.3.1 and List 4.3.2).

LIST 4.3.1. PERL SCRIPT `indexer.pl` **CREATES AN INDEX**

```
#!/usr/local/bin/perl
open (TEXT, "1DFRE10.TXT")||die"cannot";
open (OUT, ">1DFRE10.OUT")||die"cannot";
$line = " ";
@indextermarray = ("gaul","roman
    empire","emperor","village","england");
while ($line ne "")
    {
    $cumline = "";
    for($i=0;$i<100;$i++) #assumes 100 lines per page
        {
        $line = <TEXT>;
        $cumline = $cumline . $line;
        }
    $page++;
    $cumline = lc($cumline);
    $cumline =~ s/[^a-z\-\']/ /g;
    $cumline =~ s/ +/ /g;
```

```
    foreach $thing (@indextermarray)
        {
        if ($cumline =~ /\b$thing\b/)
            {
            $wordpage{$thing} = $wordpage{$thing} . " $page";
            }
        }
    }
#The index is finished.
#The next lines just display it on-screen.
foreach $key (sort keys %wordpage)
    {
    print OUT "$key \= $wordpage{$key}\n";
    print "$key \= $wordpage{$key}\n";
    }
exit;
```

LIST 4.3.2. AN EXCERPTED OUTPUT FOR THE INDEXING PROGRAM, LISTING ONLY THE TERMS "ENGLAND" AND "VILLAGE" AND THE PAGES ON WHICH THEY ARE FOUND

```
c:\ftp>perl indexer.pl
.
.
.
england = 5 14 23 134 207 208 229 277
.
.
.
village = 43 77 81 94 128 141 147 184 185 225 226 244
```

4.4. AUTOMATIC TERM EXTRACTION

Professional indexers create most indexes. Indexers read through a text, note the phrases that might be of particular interest to readers, and jot down the page number where the phrase was found. There are three problems with this approach (List 4.4.1).

It would be helpful to have an automated way to select index terms. Kim and Wilbur wrote a thoughtful paper in which they describe six automatic methods for extracting candidate phrases-of-interest from text (4).

LIST 4.4.1. PROBLEMS WITH HUMAN-BASED INDEXING

Labor-intensive and time consuming. The index cannot be built until the book is in final form and the page numbers are known, thus delaying publication of the book until the indexing is completed.

If important phrases are omitted completely or if one or more of the phrase locations are omitted, it is likely that no one will catch the error.

The indexing effort must be repeated if there are book revisions and pagination changes.

The simplest method selects all commonly occurring two- or three-word phrases that lie between so-called stop words in the text. Stop words are short, frequently occurring words such as "a, and, but, if, when" that almost never occur within index-worthy phrases. Stop words serve as barriers enclosing candidate phrases (List 4.4.2 and List 4.4.3).

LIST 4.4.2. PERL SCRIPT EXTRACTS CANDIDATE INDEX PHRASES FROM A FILE

```
#!/usr/local/bin/perl
@stop = qw(
a about absent absence again all almost also although always
   among an
and another any are as at be because been before being
   between both
but by can could cm did do does done due during each either
   enough
especially etc for found from further had has have having
   here how
however i if in into is it its itself just kg km made mainly
   make may
mg might ml mm most mostly must nearly neither no nor observed
obtained often on our overall perhaps present presence quite
   rather
really regarding seem seen several should show showed shown
   shows
significantly since so some such than that the their theirs
   them then
there therefore these they this those through thus to upon
   use used
using various very was we were what when which while with within
without would or can't doesn't not
);
```

```
open (TEXT,"1DFRE10.TXT")||die"Cannot";
open (OUT,">1DFRE10.OUT")||die"Cannot";
undef($/);
$phrase = <TEXT>;
$phrase =~ s/\n/ /g;
$phrase = lc($phrase);
$phrase =~ s/[^a-z \']/ /g;
foreach $stopword (@stop)
  {
  $phrase =~ s/ $stopword / \# /g;
  }
$phrase =~ s/[\s]+/ /g;
$phrase =~ s/ ?\# ?/\#/g;
@phraselist = sort (split("#",$phrase));
@phraselist = grep
{$i{$_}++;(($i{$_}==2)&&(scalar(split(" ",$_))>1));}@phraselist;
print OUT join("\n",@phraselist);
exit;
```

LIST 4.4.3. FIRST NINE LINES OF OUTPUT FROM THE `phrase.pl` SCRIPT

abate fortis
abbe foucher
abdication of diocletian
abilities of
able leader
abolition of
absolute power
abuse of
academy of inscriptions

4.5. BRAIN-TEASER SECTION, THE BWT (BURROWS-WHEELER TRANSFORM)

This section is included for readers with an insatiable appetite for complex indexing algorithms. The casual reader may skip this section and return to it if they ever encounter a situation wherein the size of their datasets combined with their indexes becomes excessively large.

Indexes are a fast method of finding and retrieving text, but they come with an annoying problem: good indexes are large. A complete index can easily exceed the length of the indexed text. Storing, organizing, and accessing a file and its index can be difficult. Imagine how nice it would be if we could somehow transform text files into a magical format that combined the text and the index in a single file that was smaller than the text itself! Imagine further that the transformed file would allow us to read the file, starting with any word contained in the file, without needing to read sequentially through the file up to the point of the chosen start-word.

The Burrows-Wheeler transform (BWT), invented in 1994, can do all this and more (5). BWT and its related suffix arrays are used in a popular open-source compression algorithm (6), as an indexing algorithm (7), and as an annotation algorithm for large genome databases (8).

BWT converts a block of text into a list of words that are ordered in a special sequence that can be transformed into the original block of text, using a derived transformation vector. Because a transformation vector points any text word to its preceding and succeeding words, it can be used as a complete index to the text.

We will use the sentence, "I will be here soon," to show how the Burrows-Wheeler transform builds the transformation vector and reconstitutes the sentence from the vector.

Step 1: Build an array of the consecutive circular forms of the sentence created by moving the first word to the end of the sentence.

```
0        I        will     be       here     soon
1        will     be       here     soon     I
2        be       here     soon     I        will
3        here     soon     I        will     be
4        soon     I        will     be       here
```

Step 2: Sort alphabetically on the first column to yield columns f(irst) and l(ast).

```
         f(irst)                             l(ast)
2->      be       here     soon     I        will
3->      here     soon     I        will     be
0->      I        will     be       here     soon
4->      soon     I        will     be       here
1->      will     be       here     soon     I
```

Step 3: Sort in place alphabetically on the last column.

```
         f(irst)  l(ast)
2->      be       will     -> 4
3->      here     be       -> 0
0->      I        soon     -> 3
4->      soon     here     -> 1
1->      will     I        -> 2
```

Step 4. Compute the transformation vector that connects each element in the f column to its succeeding word in the l column.

```
          T(ransform)
          vector
2 -> 1    "be" ->  "here"
3 -> 3    "here" -> "soon"
0 -> 4    "I" -> "will"
4 -> 2    "soon" is "I"
1 -> 0    "will" -> "be"
```

The transformation vector, T, is:

```
0 1 2 3 4 element number

1,3,4,2,0 vector element
```

Step 5. Reverse the transformation.

The transformation vector is:

```
[0-1, 1-3, 3-2, 2-4, 4-0]
```

The L column is:

```
0      1     2      3      4      f column word number
will,  be,   soon,  here,  I      f column word
```

The transformation vector can be used to compute the original sentence from the f column:

"be here soon I will", with transform "be(0-1), here (1-3), soon (3-2), I (2-4), will (4-0)."

Wait! There is something wrong here. The sentence is present, in the correct word order (for a circular sentence), but it begins with the second word, "be," instead of the 0th word, "I."

The BWT does not care where a text begins or ends. That's why it can be used so well for indexing and retrieval. If you want to reconstitute text in its original order, you must keep track of the first word in the text (the so-called primary index of the BWT).

After this strange exercise, can we assess what we have done? Yes. The BWT consists entirely of column l (the last column determined in Step 2), and the primary index (the place where the text starts). Everything else (column f, the transform vector, and the text itself) can be generated from column l and the primary index. Remember that column f is the ordered list of words in the text. Column l is a strangely ordered list of words in the text. You can generate column f from column l simply by alphabetically ordering the l column. We have already seen that the transform vector can be generated from columns f and l, and we have seen that the text can be reconstituted from the transform vector and the l column. So that's really everything!

The BWT is an incredibly powerful tool. The f column (first column) is the ordered list of words from the text. It provides an automatic count of all the different words in the text and their frequencies. The BWT provides the index (location) for each word in the file. A clever informaticist might find many ways to use the BWT in search-and-retrieval algorithms and in data-merging projects. Only two tasks remain:

1. Stand in awe of Burrows and Wheeler for creating this marvelous algorithm.
2. Implement the algorithm in Perl.

The Perl script `bwt.pl` is a simple version of the BWT algorithm. Because it does not carve the input text into blocks, it is limited by the size of the computer's memory (the text transformation arrays can become very large). A sample text (List 4.5.1) and the Perl script (List 4.5.2) is shown. For simplicity, the text is lowercase and without punctuation.

The heart of the script involves computing the transform vector (`@T`, actually the transform array) and reversing the transform. Compare the following lines to Steps 4 and 5:

```
for(my $i=0;$i<$arraylength;$i++)
  {
  for (my $n=0;$n<$arraylength;$n++)
    {
    if ($f[$i] eq $l[$n])
       {
       push(@T,$n);
       $l[$n]="";
       last;
       }
    }
  }
foreach $transform (@T)
   {
   $transformcount{$count} = $transform;
   #transforms to the next word
   $transformtransform{$transform} = $count;
   #transforms to the prior word
   #print "$count $transform\n";
   $count++;
   }
```

A BWT file is a transformed replacement for both the original file and the file's index. Using the BWT file, you can recompose the original file, or you can find any portion of a file preceding or following any word from the file. The BWT is an amazing informatics invention.

LIST 4.5.1. SAMPLE INPUT, `text.txt`, FOR `bwt.pl`, FOLLOWED BY PARTIAL OUTPUT

```
mesh is used by the national library of medicine to index
   all biomedical abstracts included in medline and has been
   used to index medical terms found throughout the internet
   mesh is a mature well-curated large comprehensive and
   publicly available nomenclature mesh alone is a sufficient
   medical vocabulary for many indexing purposes mesh can be
   downloaded at no cost mesh is used by the national library
   of medicine to index all biomedical abstracts included in
   medline and has been used to index medical terms found
   throughout the internet mesh is a mature well-curated
   large comprehensive and publicly available nomenclature mesh
   alone is a sufficient medical vocabulary for many indexing
   purposes
```

Partial output:

(76,53) mesh
(53,103) is
(103,22) used
(22,95) by
(95,78) the
(78,57) national
(57,83) library
(83,67) of
(67,99) medicine
(99,41) to
(41,6) index
(6,20) all
(20,4) biomedical
(4,39) abstracts
(39,37) included
(37,69) in

LIST 4.5.2. PERL SCRIPT `bwt.pl` **IMPLEMENTS THE BURROWS-WHEELER TRANSFORM**

```
#!/usr/local/bin/perl -w

open (TEXT,"text.txt")||die"Cannot";
open (OUT,">bwt.txt")||die"Cannot";
my $line = " ";
$/ = undef;
$line = <TEXT>;
$/ = "\n";
$line = lc($line);
$line =~ tr/a-z/ /c;
$line =~ s/\n/ /g;
$line =~ s/^ +//g;
$line =~ s/ +$//g;
$line =~ s/ +/ /g;
@linearray = split(/ +/,$line);
$firstword = @linearray[0];
$arraylength = scalar(@linearray);
foreach my $thing (@linearray)
   {
my $thing = $thing .
   "\%\%\%\%\%\%\%\%\%\%\%\%\%\%\%\%\%";
   $thing = substr($thing, 0 , 15);
   push(@newlinearray,$thing);
   }
@linearray = @newlinearray;
my $count = 0;
while ($count < (scalar(@newlinearray)))
   {
   $bwtline = join("",@linearray);
   push(@bwtarray, $bwtline);
   $first = shift(@linearray);
   push (@linearray, $first);
   $count++;
   }


@doubletarray = sort(@bwtarray);
$count = 0;
foreach $thing (@doubletarray)
  {
  my $begin = substr($thing,0,15);
```

```
    $begin =~ s/\%//g;
    my $end = substr($thing,-15,15);
    $end =~ s/\%//g;
    print OUT "$begin $end $count\n";
    if ($end eq $firstword)
      {
      $beginplace = $count;
      undef($firstword);
      }
    push (@f, $begin);
    push (@l, $end);
    $count++;
    }
  my $count = 0;
  for(my $i=0;$i<$arraylength;$i++)
     {
     $lhash{$count}=$l[$i];
     $count++;
     }
  print OUT "\n\n";
  my @T;
  for(my $i=0;$i<$arraylength;$i++)
    {
    for (my $n=0;$n<$arraylength;$n++)
      {
      if ($f[$i] eq $l[$n])
         {
         push(@T,$n);
         $l[$n]="";
         last;
         }
      }
    }
  $count = 0;
  foreach $transform (@T)
     {
     $transformcount{$count} = $transform;
     #transforms to the next word
     $transformtransform{$transform} = $count;
     #transforms to the prior word

     $count++;
     }
```

```
print join(",", @T);
$count = 0;
while ($count < $arraylength)
    {
print OUT "$lhash{$beginplace}
    \($beginplace\,$transformcount{$beginplace}\) ";
print "$lhash{$beginplace}
    \($beginplace\,$transformcount{$beginplace}\) ";
    $beginplace = $transformcount{$beginplace};
    $count++;
    }
    exit;
```

CHAPTER

5

Autocoding Biomedical Data with Nomenclatures

5.1. BACKGROUND

Modern nomenclatures are used to organize, index, and retrieve biomedical data. Most modern nomenclatures are prepared as a taxonomy (collection of the relevant items in a data domain), wherein synonymous terms are grouped together and assigned a unique concept code.

For instance, in the Developmental Lineage Classification and Taxonomy of Neoplasms (hereinafter called the Neoplasm Classification), prostate cancer is assigned the unique concept code C486300, and all the term variants for this concept are attached to the same concept code (9), (10), (11), (12), (List 5.1.1).

LIST 5.1.1. EQUIVALENT TERMS FOR THE CONCEPT IDENTIFIER C4863000

```
C4863000  prostate with adenoca
C4863000  adenoca arising in prostate
C4863000  adenoca involving prostate
C4863000  adenoca arising from prostate
C4863000  adenoca of prostate
C4863000  adenoca of the prostate
C4863000  prostate with adenocarcinoma
C4863000  adenocarcinoma arising in prostate
C4863000  adenocarcinoma involving prostate
C4863000  adenocarcinoma arising from prostate
C4863000  adenocarcinoma of prostate
```

```
C4863000  adenocarcinoma of the prostate
C4863000  adenocarcinoma arising in the prostate
C4863000  adenocarcinoma involving the prostate
C4863000  adenocarcinoma arising from the prostate
C4863000  prostate with ca
C4863000  ca arising in prostate
C4863000  ca involving prostate
C4863000  ca arising from prostate
C4863000  ca of prostate
C4863000  ca of the prostate
C4863000  prostate with cancer
C4863000  cancer arising in prostate
C4863000  cancer involving prostate
C4863000  cancer arising from prostate
C4863000  cancer of prostate
C4863000  cancer of the prostate
C4863000  cancer arising in the prostate
C4863000  cancer involving the prostate
C4863000  cancer arising from the prostate
C4863000  prostate with carcinoma
C4863000  carcinoma arising in prostate
C4863000  carcinoma involving prostate
C4863000  carcinoma arising from prostate
C4863000  carcinoma of prostate
C4863000  carcinoma of the prostate
C4863000  carcinoma arising in the prostate
C4863000  carcinoma involving the prostate
C4863000  carcinoma arising from the prostate
C4863000  prostate adenoca
C4863000  prostate adenocarcinoma
C4863000  prostate ca
C4863000  prostate cancer
C4863000  prostate carcinoma
C4863000  prostatic cancer
C4863000  prostatic carcinoma
C4863000  prostatic adenocarcinoma
C4863000  prostate gland adenocarcinoma
C4863000  adenocarcinoma of the prostate gland
C4863000  adenocarcinoma of prostate gland
C4863000  prostate gland carcinoma
C4863000  carcinoma of the prostate gland
C4863000  carcinoma of prostate gland
```

When a nomenclature collects synonymous terms under a unique concept identifier, medical text containing any of the synonymous terms can be assigned the same unique concept code. When all the terms in a medical database have been coded with concept identifiers,, they can be retrieved through a search that collects synonymous terms through their unifying concept identifier.

Medical autocoding can be considered a specialized form of machine translation (13). Machine translation is a large field that covers direct translations between different languages (for example, Russian to and from English), the interconversion of language modes (for example, audible words to and from text), the interpretation of signals (for example, military SIGINT, deriving intelligence from the analysis of intercepted signals), the annotation of text through the extraction of terms and concepts (that is, medical autocoding), and the transformation of text into desired data structures (that is, converting narrative text to tagged XML).

There are several computational approaches to machine translation. The first is to parse sentences into grammatical tokens permitting a program to re-order component parts of the sentence into a sequence of phrases that make grammatical sense in the target language. The sequential phrases of the transformed sentence can then be matched against a controlled vocabulary (in the target language) yielding translated text. The problem with this approach is that it is computationally intensive (resulting in slow execution speed) and error-prone when sentences are long, complex, or idiomatic. Machine translation depends on the creation of elaborate grammar rule systems and exception lists that account for commonly used non-grammatic language structures.

A second approach is "lexical" parsing, in which text is parsed and any phrases in the text that exactly match terms held in a nomenclature are extracted. The lexical parser does not tokenize a sentence into grammatical parts and does not re-order component parts of a sentence into alternate forms of statements. Lexical parsers are a simple but somewhat brutish approach to machine translation. The lexical parser depends on terms existing in text without splitting modifiers. For instance, a lexical parser would easily extract the term "flat feet" when it appears in a medical text and is included in a medical nomenclature. If the phrase "flat erythemic feet" appeared in a medical text, the lexical parser would miss the term, unless flat erythemic feet were included in the nomenclature (as a specially denoted form of flat feet).

As used in this manuscript, the term "autocoder" refers to a software program capable of parsing large collections of medical records (for example, radiology reports, surgical pathology reports, autopsy reports, admission notes, discharge notes, operating room notes, medical administrative e-mails, memoranda, and manuscripts), capturing the medical concepts contained in the text and assigning them an identifying code from a nomenclature.

The term "autocoding" should be distinguished from "computer-assisted manual coding." Health-care workers may use a software enhancement of their Hospital Information Systems to code a section of text as they enter reports into the computer system. Typically, candidate terms and term codes are displayed on the same screen as the entered report. The person entering text is often given the option of editing the proffered codes. This process should not be confused with "autocoding" and is not equivalent to the fully automatic and large-scale coding required in data-mining projects, database creation, and large-scale quality reviews.

Finding all the concepts in a corpus of text is a necessary and early step in all data-mining efforts. The autocoded terms can be used individually as index terms for the document, on a record-by-record basis to produce a concept "signature" that is highly specific for each report, or collectively to relate the frequency of terms within records with the frequency of terms in the aggregate document (14).

The purpose of this chapter is to introduce the reader to some of the freely accessible nomenclatures in biomedicine and to provide some techniques for autocoding biomedical text.

5.2. DOUBLET ALGORITHM FOR A FAST LEXICAL AUTOCODER

The doublet method is a novel approach to autocoding. It can autocode several megabytes of text per second on standard desktop computers (circa 2007). This section describes the doublet method algorithm.

One of the many problems in the field of machine translation is that expressions (multiword terms) convey ideas that transcend the meanings of the individual words in the expression. Consider the following sentence:

"The ciliary body produces aqueous humor."

The example sentence has unambiguous meaning to anatomists, but each word in the sentence can have many different meanings. "Ciliary" is a common medical word, and usually refers to the action of cilia. Cilia are found throughout the respiratory and GI tract and have an important role locomoting particulate matter. The word "body" almost always refers to the human body. The term "ciliary body" should (but does not) refer to the action of cilia that move human bodies from place to place. The word "aqueous" always refers to water. Humor relates to something being funny. The term "aqueous humor" should (but does not) relate to something that is funny by virtue of its use of water (as in squirting someone in the face with a trick flower). Actually, "ciliary body" and "aqueous humor" are each examples of medical doublets whose meanings are specific and contextually constant (that is, always mean one thing). Furthermore, the meanings of the doublets cannot be reliably determined from the individual words that constitute the doublet, because the individual words have several different meanings. Basically, you either know the correct meaning of the doublet, or you don't.

Any sentence can be examined by parsing it into an array of intercalated doublets:

> "The ciliary, ciliary body, body produces, produces aqueous,
>
> aqueous humor."

The important concepts in the sentence are contained in two doublets (ciliary body and aqueous humor). A nomenclature containing these doublets would allow us to extract and index these two medical concepts. A nomenclature consisting of single words might miss the contextual meaning of the doublets.

What if the term were larger than a doublet? Consider the tumor "orbital alveolar rhabdomyosarcoma." The individual words can be misleading. This orbital tumor is not from outer space, and the alveolar tumor is not from the lung. The three-word term describes a sarcoma arising from the orbit of the eye that has a morphology characterized by tiny spaces of a size and shape as might occur in glands (alveoli). The term "orbital alveolar rhabdomyosarcoma" can be parsed as "orbital alveolar, alveolar rhabdomyosarcoma." Why is this any better than parsing the term into individual words, as in "orbital, alveolar, rhabdomyosarcoma"? The doublets, unlike the single words, are highly specific terms that are unlikely to occur in association with more than a few specific concepts.

Very few medical terms are single words. In the developmental lineage classification of neoplasms, there are over 100,000 unique terms for neoplasms (9). All but 252 of these terms are multiword terms.

Several innovative approaches to autocoding have used the higher information content of multiword terms (also called word n-grams) to match terms in text with terms in vocabularies or to enhance the content of vocabularies by identifying n-grams occurring in text that qualify as new nomenclature terms (15).

The doublet method uses the higher term specificity of doublets [bi-grams] to construct a simple and fast lexical parser. Lexical parsers are types of string-matching algorithms. In general, the overall speed of lexical parsers is determined by the speed with which the parser can prepare an array of all possible words and phrases contained in a block of text, coupled with the speed with which each of these phrases can be compared against all the terms in the nomenclature.

The algorithm for the doublet autocoder is shown in the following (List 5.2.1).

The Perl script implementing the doublet algorithm has been placed in the public domain and can be downloaded freely as a supplemental file attached to its original publication (16). It is too long to include in its entirety here, but we will implement part of the algorithm in the next section, and we will see the full algorithm implemented in Chapter 19.6.

LIST 5.2.1. ALGORITHM FOR THE DOUBLET AUTOCODER

`doubcode.pl`

1. Each phrase (term) in the nomenclature (`neocl.xml`) is converted into intercalated doublets, and each doublet is assigned a consecutive number.
2. Each nomenclature phrase is assigned the concatenated list of numbers that represent the ordered doublets composing the phrase.
3. Every text record (PubMed abstract, in this case) is split into an array consisting of the consecutive words in the text record.
4. The text array is parsed as intercalated doublets. Intercalated doublets from the text that match doublets found anywhere in the nomenclature are assigned their numeric values (from the doublet index created for the nomenclature). Runs of consecutive doublets from the text that match doublets from the nomenclature are built into concatenated strings of doublet values. The occurrence of a text doublet that does not match any doublet in the nomenclature cannot possibly be part of a nomenclature term. Such text doublets serve as "stop" doublets between candidate runs of text doublets that match nomenclature doublets.
5. The runs of matching doublets are tested to see if they match any of the runs of doublets that compose nomenclature terms, or if they contain any subsumed terms that match nomenclature terms.
6. The array of doublet runs extracted from the text that match nomenclature terms are cached.

5.3. AUTOMATIC EXPANSION OF A MEDICAL NOMENCLATURE

The doublet method has numerous algorithmic variants, one of which helps curators find new nomenclature terms from the medical literature.

Nomenclatures are not static documents. New medical terms appear constantly. The best way to add terms to an existing vocabulary is to read the current literature in the knowledge domain of the nomenclature, and transcribe new terms when they are encountered. It is difficult to imagine any automatic process that can replace this scholarly pursuit. Terms encountered while reading a scientific text appear in a structured context that often defines the term, clarifies the relationship of the new term to related terms, and sometimes provides sufficient information to classify the new term within a structured taxonomic hierarchy.

It is unfortunate that it is impossible for curators to read all of the biomedical literature pertaining to a nomenclature's domain. An algorithm described in List 5.3.1 parses through a large corpus of text and extracts phrases from the text that are not included in the current nomenclature (that is, are not known nomenclature terms) but which have a feature suggesting that the phrase might be a new nomenclature term. Specifically, the algorithm finds phrases composed of word doublets that are present somewhere within one or more terms of the nomenclature. Text phrases composed of sequences of word doublets found in an existing nomenclature are candidates for new nomenclature terms (List 5.3.1). The rationale for the algorithm is that most nomenclature terms are composed of word doublets found in other nomenclature terms. This general method can be used with any text and any existing nomenclature. This method permits curators to continually enhance their nomenclatures with new terms, an essential activity needed to ensure the proper coding and annotation of biomedical data.

LIST 5.3.1. DOUBLET METHOD FOR FINDING CANDIDATE TERMS FROM TEXT

1. Collect all the doublets that occur in the entire nomenclature (that is, accumulate a list of the doublets from every term in the nomenclature).
2. Parse a text file into an ordered collection of overlapping doublets. As an example, the text "serous borderline ovarian tumor" would be parsed as "serous borderline, borderline ovarian, ovarian tumor."
3. Compare each consecutive text doublet against the array of doublets from the nomenclature to determine whether the doublet exists somewhere in the nomenclature.

4. If the doublet from the text does not exist in the nomenclature, it can be deleted. If it exists in the nomenclature, it is concatenated with the next text doublet if the next text doublet exists in the nomenclature. Otherwise, it is deleted. This process continues, concatenating doublets that exist somewhere in the nomenclature. Extraneous leading words (the, in, of, with, and) and trailer words (the, and, with, from, a) are automatically deleted from the final concatenated sequence. Final concatenated sequences of two or greater consecutive doublets that match to doublets from the nomenclature are saved as candidate terms.

The complete Perl implementation `getdoub.pl` is too long to include in this book. It is readily available for download as a public domain file attached to the original article describing the algorithm (17). A Perl snippet suffices to demonstrate the most functional segments of the code (List 5.3.2).

The snippet of Perl shows the heart of the term-extraction algorithm.

LIST 5.3.2. PERL SNIPPET FROM `getdoub.pl` TO EXTRACT CANDIDATE TERMS FROM A NOMENCLATURE

```
.....
@hoparray = split(/ /,$line);
#make an array from the words in the text
my $olddoublet = "";
for ($i=0;$i<(scalar(@hoparray)-1);$i++)
   {
   $doublet = "$hoparray[$i]$hoparray[$i+1]";
   #iterate by doublets through the word array
   if (exists $doubhash{$doublet})
   #if the doublet exists somewhere in the nomenclature
       {
       if ($englishline ne "")
         {
         $englishline = $englishline . " $hoparray[$i+1]";
         #extend the candidate phrase with the next word
         }
       else
         {
         $englishline = $doublet;
         }
       }
```

At the top of the snippet, a fragment of unpunctuated text is split at the space character between words, and the resulting list of consecutive words from the text is assigned to an array:

```
@hoparray = split(/ /,$line);
```

For the number of words in the array, each overlapping word doublet is assigned to a variable through a for-loop:

```
for ($i=0;$i<(scalar(@hoparray)-1);$i++)
   {
$doublet = "$hoparray[$i]$hoparray[$i+1]";
```

An associative array of each doublet occurring in a reference nomenclature was prepared in a prior segment of the script, and this associative array has been assigned to `%doubhash`. Each doublet occurring in `@hoparray` is interrogated to determine if the doublet exists in `%doubhash`:

```
if (exists $doubhash{$doublet})
```

If the doublet exists in `%doubhash`, the second word of the doublet is added to the concatenation of consecutive doublets occurring in `@hoparray` and found in the nomenclature (*`$englishline`*):

```
{
if ($englishline ne "")
    {
    $englishline = $englishline . " $hoparray[$i+1]";
    }
```

If the concatenation of consecutive doublets is empty, the doublet starts the new string:

```
 else
    {
    $englishline = $doublet;
    }
 }
```

The purpose of the doublet phrase extractor is to parse through any corpus of text, extracting phrases that might contain new nomenclature terms. The phrases are chosen to meet two criteria (List 5.3.3).

The doublet extractor works fast to produce a neat list of candidate phrases that can be conveniently reviewed by a curator. In a test case, a 31+ megabyte corpus was extracted in 2 seconds, to produce 313 candidate phrases (17). From the candidate phrases, the curator selected 222 phrases that could be added to the reference nomenclature (requiring 30 minutes of the curator's time.).

LIST 5.3.3. CRITERIA FOR INCLUDING A PHRASE AS A CANDIDATE NEW TERM

1. Candidate phrases extracted from a text corpus are composed of concatenated strings of word doublets that match doublets found in an existing nomenclature.
2. Candidate phrases are excluded if they already occur in the nomenclature.

5.4. EXTRACTING THE CATEGORY 0 VOCABULARIES FROM UMLS

The UMLS Metathesaurus is the largest curated medical nomenclature in existence. It is composed of more than 100 different biomedical vocabularies and contains about 6 million term records.

The UMLS files are available at no cost from the National Library of Medicine's Web site (see Appendix). However, there are usage restrictions attached to many of the contained vocabularies. Vocabularies with no special usage restrictions are often referred to as the Category 0 vocabularies. Vocabularies contributed by the U.S. government, or that were created substantially with U.S. government funds, form the bulk of the Category 0 vocabularies. Scientists can annotate their data files with UMLS codes taken from the Category 0 vocabularies and can distribute their datafiles to their colleagues or to the public (18).

The non-Category 0 vocabularies encumber UMLS users with restrictions that may prohibit: 1) distribution to colleagues, 2) posting on a publicly available site (such as an Internet Web site), 3) submission to shared data-set repositories, or submitted as supplemental data in support of research articles, and 4) submission to scientific journals.

Consequently, professionals who use the UMLS might feel the need to extract the Category 0 vocabularies.

MRCONSO is the UMLS metathesaurus file that contains every term record. Each term record is a line in the MRCONSO flat file. Let us look at the first record in MRCONSO:

C0000005|ENG|X|L0000005|PF|S0007492|Y|A7755565||M0019694|

D012711|MSH|PEN|D012711|(131)I-Macroaggregated Albumin|0|N||

The UMLS unique concept code is "C0000005". The term is "(131)I-Macroaggregated Albumin".

The data elements of each record are separated by a vertical line character, "|". The zero place element is the UMLS concept unique identifier code (C0000005 in the example). The first element of each MRCONSO record is the language for the term (ENG for English in the example). The eleventh data element is the name of the source vocabulary (MSH for MESH in this example). The fourteenth data element is the term name, (131)I-Macroaggregated Albumin. The 700+ megabyte MRCONSO file contains millions of unique term records. A single concept identifier code might apply to many different synonymous terms.

We can use our knowledge of the MRCONSO record format to determine the exact number of term records and the number of unique concepts contained in the UMLS Metathesaurus (List 5.4.1 and List 5.4.2).

It is simple to write an extraction program to parse each line of text, pulling out concepts and terms for any particular source vocabulary. List 5.4.3. is a script that extracts every SNOMED-CT vocabulary term from MRCONSO.

LIST 5.4.1. PERL SCRIPT `terms.pl` COUNTS THE TERMS AND CONCEPTS IN MRCONSO

```
#!/usr/local/bin/perl
open (TEXT, "mrconso")||die"cannot";
$line = " ";
$count = 0;
while ($line ne "")
  {
  $count++;
  $line = <TEXT>;
  @linearray = split(/\|/,$line);
  $code = $linearray[0];
  next if (exists($codehash{$code}));
  $codehash{$code}="";
  }
print "Term count is $count\n\n";
$count = 0;
while (($key,$value) = each(%codehash))
   {
   $count++;
   }
print "Concept count is $count";
end;
```

LIST 5.4.2. OUTPUT OF `terms.pl`

```
c:\ftp>perl terms.pl

Term count is 6405469

Concept count is 1352404
```

LIST 5.4.3. PERL SCRIPT `vocab.pl` **EXTRACTS SNOMED-CT TERMS FROM MRCONSO**

```
#!/usr/bin/perl
open (TEXT,"c\:\\ftp\\entrez\\mrconso");
#contains location on my computer for MRCONSO file
$line = " ";
open (OUT,">snom.txt");
while ($line ne "")
    {
    $line = <TEXT>;
    @linearray = split(/\|/,$line);
    $cuinumber = $linearray[0];
    $language = $linearray[1];
    $vocabulary = $linearray[11];
    next if ("ENG" ne $language);
    next unless ($vocabulary =~ /SNOMEDCT/);
    print OUT $line;
    }
exit;
```

5.5. COLLECTING THE ICD CODES FROM THE UMLS METATHESAURUS

Of course, any individual vocabulary can be easily extracted from the UMLS metathesaurus if you know the name of the vocabulary as used by UMLS. List 5.5.1 extracts the ICD (International Classification of Disease) vocabulary terms from MRCONSO, reformatting each record to consist of the Concept Unique Identifier Code followed by the vocabulary term.

We will use the `icd.txt` file, produced here, again in Chapter 13. The script can easily be modified to extract any and all of the Category 0 vocabulary terms from the UMLS Metathesaurus. The names of Category 0 vocabularies, as they may appear in element 11 of the MRCONSO record, are: “AIR, AOD, CCS, COSTAR, CSP, DXP, HCPCS, HL7, ICD9CM, ICPC2P, ICPCBAQ, ICPCDAN, ICPCDUT, ICPCFIN, ICPCFRE, ICPCGER, ICPCHEB, ICPCHUN, ICPCITA, ICPCNOR, ICPCPOR, ICPCSPA, ICPCSWE, LCH, LNC, MCM, MSH, MTH, MTHHH, MTHICD9, MTHMST, MTH, MSTFRE, MTHMSTITA, NCBI, NCI, PDQ, QMR, RAM, RXNORM, SPN, SRC, UWDA, and VANDF.

LIST 5.5.1. PERL SCRIPT `ICD.pl`, FOR COLLECTING ICD CODES FROM THE UMLS METATHESAURUS

```
#!/usr/bin/perl
$line = " ";
$start = time();
open (TEXT,"c\:\\ftp\\entrez\\MRCONSO");
#my own computer's path to the UMLS MRCONSO file

open (OUT,">icd.txt");
while ($line ne "")
   {
   $line = <TEXT>;
   @linearray = split(/\|/,$line);
   $icdnumber = $linearray[13];
   $language = $linearray[1];
   $term = $linearray[14];
   $vocabulary = $linearray[11];
   next if ("ENG" ne $language);
   next unless ($vocabulary =~ /ICD10AM/);
   print OUT "$icdnumber $term\n";
   }
$end = time();
$total = $end - $start;
print "\ntotal time was $total seconds\n";
exit;
```

CHAPTER 6

Searching and Mining Data

6.1. BACKGROUND

There are some questions that cannot be answered by experience, by reviewing the literature, or by experiment. Some questions can be answered only by reviewing all of the collected information in a knowledge domain and finding a constant trend, a class of cases, or a shared property that reveal some new biomedical truth.

The typical approach to data searching assumes that two basic conditions will always hold:

1. The searcher is expected to know what she is searching for.
2. The collection of data is expected to be organized in a manner that will ensure that an item will be found if it is contained in the data collection.

Except for the most trivial situations, those two conditions almost never apply. New searches tend to start off as unrefined queries. As query results are obtained, the searcher gets an idea of the range of subject information available and modifies the query until a final search item is chosen. The process of query refinement is crucial. The search engine must provide the searchers with a valid notion of the depth and breadth of the contained information. The search engine must help the searcher determine what she is searching for.

The prototypical biomedical data search is a physician seeking a patient's test results (for example, a biopsy, a blood analysis, an x-ray). Beyond simple searches for the medical records of individual patients, hospital information systems are rather poor data mining tools.

Consider a very simple database search. The hospital nutritionist may wish to query the hospital information system for a list of all patients who are overweight. This search would be problematic unless patient weight was recorded as a database field and the data field was linked to a date. Most hospital information systems have no "weight" element in their data records. In some systems, records in the hospital information system are primarily billable procedures (surgical operations, radiology reports, pharmacy data). An observation, such as the patient's weight on a certain date, is unlikely to be available for query. Even if this information were available in the database, the information system may have no facility for collecting data from the entire population of registered hospital patients. And if the hospital information system were capable of querying for multiple patients, how would the query determine which patients are overweight? A middle-aged woman of height 5 feet 5 inches may be underweight if she weighs 100 pounds. A 4-year-old boy who weighs 100 pounds would be overweight. A medical report might be entered into a database as a record containing a patient name field, a date field, and a procedure field, with the text of the report attached as a "string" object. In such a case, the database may accept queries on patient name and date and procedure, but may not accept queries on items contained in text strings attached to the report. If you're uncertain what you're looking for until you see it, and if you need to have access to all the information, structured and unstructured, in the dataset, your hospital information systems will likely disappoint you.

In practice, biomedical professionals have limited query access to data embedded within the text of reports. Researchers and healthcare professionals need to learn some fundamental data mining skills if they wish to use the data contained in hospital information systems.

Data mining is a special kind of data searching in which a valued class of data is sought from within one or more data sets. The term "data mining" implies that there is some precious and valuable substance hidden within datasets. This is seldom the case. Most biomedical data has no particular intrinsic meaning. The role of the biomedical data miner is to confer sense upon an otherwise inchoate data collection (19).

People like to think that data mining has become an automatic process conducted with the help of sophisticated and expensive software applications. In most cases, the value in biomedical data sets is achieved through a creative process that involves intense study leading to the discovery of generalizable trends and patterns.

In my experience, there is no substitute for writing your own search software. Datasets and queries and projects vary to such an extent that off-the-shelf solutions tend to fail.

The purpose of this chapter is to provide practical approaches to finding data in large biomedical data sets.

6.2. SEARCHING LARGE TEXT FILES

A common task that can be easily solved with a few lines of Perl involves searching a large text file for occurrences of a query submitted as a regular expression. This next script uses everything you have learned so far and provides another example of the utility of Perl regex (List 6.2.1, List 6.2.2, and List 6.2.3).

LIST 6.2.1. IMPLEMENTATION FOR REGULAR EXPRESSION SEARCHES OF TEXT FILES

1. Asks you for a regular expression to search a file. If you are not adept at regular expressions, just enter any word. Remember, a word or phrase is always the simplest regular expression.
2. Exits the script if you enter the return key without entering a regular expression.
3. Tells Perl to provide the current epoch time (number of seconds passed since some point in history).
4. Opens an enormous publicly available file (>700 Mbytes) named MRCONSO (see Appendix).
5. Reads every line of MRCONSO (and there are millions of them), testing each line to see if it contains a substring that matches the regular expression that you provided (Step 1).
6. If it finds a match, it adds the line number and the line to an external file named `regexout.txt`.
7. When it is finished reading the file, it asks Perl again for the epoch time, and determines the script execution time by subtracting the script's end time from the script's beginning time.
8. It prints to the monitor the time spent executing the script, as well as the filename containing the output of all the lines from the MRCONSO file that matched your provided regular expression.

To find the records matching an expression, Perl must open the file and sequentially examine every line in the file. In the example expression, ("adenoca.+gland"), Perl needed to find matches to the string "adenoca" followed by the string "gland", allowing for any number of intervening characters of any value (see List 6.2.3). It is lucky that Perl is very fast. Perl searched a 775 megabyte file (a file the size of about 775 novels!) on a 2.5 MHz computer in just 46 seconds!

LIST 6.2.2. PERL SCRIPT `perlfind.pl` **FOR REGULAR EXPRESSION SEARCHES OF TEXT FILES**

```
#!/usr/bin/perl
open (OUT, ">regexout.txt")||die"Can't open file $value";
$filename = "regexout\.txt";
 print "What's your search regex?\n";
 $regex = <STDIN>;
 $regex =~ s/\n//o;
 #removes newline character from user's input
 if ($regex eq "")
   {
   close TEXT;
   close OUT;
   print "\nYou didn't give a regex...Goodby\n";
   }

 $start = time();
 &searchsub;
 $end = time() - $start;
 print "Retrieval time is $end seconds.\n";
 print "Your search results are in file $filename.";


 sub searchsub
 {
 open (TEXT, "c\:\\umls\\mrconso")||die"Can't open file $value";
 $line = " ";
 while ($line ne "")
   {
   $line = <TEXT>;
   if ($line =~ /$regex/oi)
     {
     print OUT $. . "\|" . $line;
     }
   }
 }
 exit;
```

LIST 6.2.3. OUTPUT OF REGULAR EXPRESSION SEARCH

```
C:\ftp>perl perlfind.pl
What's your search regex?
adenoca.+gland
Retrieval time is 46 seconds.
Your search results are in file regexout.txt.
```

The first two lines of the output file, regexout.txt are:

1498803|C0205641|ENG|S|L3552384|VO|S4174467|Y|A7623597||C3678||NCI|SY|C3678|Basal Cell Adenocarcinoma of Salivary Gland|0|N|256|

1498804|C0205641|ENG|S|L3552384|VO|S4174478|Y|A7623608||C3678||NCI|SY|C3678|Basal Cell Adenocarcinoma of the Salivary Gland|0|N|256|

6.3. PROXIMITY SEARCHING

Often, a researcher is interested in the co-occurrence of particular words and phrases within a fragment of text. For instance, a healthcare worker might be interested in the relationship of alcohol abuse and pancreatic disease. There is a large literature on alcohol abuse and another large literature on pancreatic disease. To get both, she might want to do a search that collects text containing co-occurrences of the words "alcohol" and "pancreas," or maybe "alcohol" and the leader string "pancrea," which would match pancreas and pancreatic. She might suspect that if the terms are closely related, they should appear in close proximity, maybe within 50 characters of each other. How can she search through a large text, collecting occurrences of words within a provided proximity of each other?

A Perl script can be easily written to implement the following algorithm (List 6.3.1 and List 6.3.2).

The typical proximity search with `proxfind.pl` may look like this:

```
C:\ftp>perl proxfind.pl
Enter two search words and then the desired
proximity distance between them, with
one space between each entered item

listen voice 30
```

LIST 6.3.1. ALGORITHM FOR A PROXIMITY WORD SEARCH

1. Asks you to enter two words (your co-occurrence pair) and a number representing the distance permitted between the words.
2. If you enter the return key without entering two words followed by a number, all separated by a space, the script promptly, and without explanation, exits. (Readers can implement their own error messages.)
3. Opens text file.
4. Reads every line of file, testing each line to see if it contains the first of the search words you provided in Step 1.
5. If it finds a match, it asks Perl for the byte location of the end of the current line, and then backs up from that position the length of the line plus the proximity distance, and then moves forward the position of the match position in the line of the first search word. This puts the current position in the file to a byte location that precedes the start of the matching search word by the proximity distance.
6. It then puts into a variable all of the text starting from the current position in the file and moving forward twice the proximity distance plus the length of the search word. This creates a string variable that consists of the search word plus all the characters occurring on either side of the search word up to a distance equal to the proximity parameter.
7. It then determines if the second search word is contained in this string variable.
8. If both search words are contained in the string variable, it prints the full string variable to the screen.
9. It repeats this process until the search file is exhausted.

LIST 6.3.2. PERL SCRIPT `proxfind.pl` PERFORMS SIMPLE PROXIMITY SEARCH THROUGH A FILE

```
#!/usr/local/bin/perl
print "Enter two search words and then the desired\n";
print "proximity distance between them, with\n";
print "one space between each entered item\n\n";
$arguments = <STDIN>;
$arguments =~ s/\n//o;
@parameters = split (/ +/,$arguments);
$regex1 = $parameters[0];
exit if ($regex1 eq "");
```

```
    $regex2 = $parameters[1];
    exit if ($regex2 eq "");
    $distance = $parameters[2];
    exit if ($distance < 1);
    open (TEXT, "1DFRE10.TXT")||die"Cannot";
    #use any plain-text file
    $getline = " ";
    while ($getline ne "")
       {
       $getline = <TEXT>;
       $line = $getline;
       $line = $line . "\n";
       if ($line =~ /$regex1/oi)
          {
          $line_length = length($line);
          $line_offset = length($`);
          $place = tell(TEXT);
          $length_word = length($regex1);
          seek(TEXT,0,0);
seek(TEXT, ($place - $line_length - $distance + $line_offset + 1), 0);
          read(TEXT,$line, ($distance + $length_word + $distance));
          next if ($line !~ /$regex2/moi);
          print "\-\-\-\n$line\n";
          }
       }
    exit;
```

The results of the search will appear instantly (with a 1 Mbyte file), such as *The History of The Decline and Fall of the Roman Empire*.

```
—-ancient and opulent family, listened to the voice of ambi-
tion; an
—-e might imagine that we were listening to the voice of
Moses, when
```

6.4. FINDING NEEDLES FAST USING A BINARY-TREE SEARCH OF THE HAYSTACK

It is easy for Perl to find a specific item in an ordered list, no matter how long the list. If the list is ordered, it can be gigabytes in length and Perl can find any record almost instantly.

LIST 6.4.1. PERL SCRIPT `find_bin.pl` PERFORMS A BINARY SEARCH ON A FILE

```
#!/usr/bin/local/perl
open (TEXT, "find_bin.txt");
seek(TEXT, 0, 2);
print "What word would you like to find?\n";
$findword = <STDIN>;
$findword =~ s/\n$//o;
$filesize = tell (TEXT);
for($i=1;$i<129;$i++)
   {
   $portion = int(($filesize * $i)/128);
   push(@portionarray,$portion);
   }
seek(TEXT, 0, 0);
$arraynumber = 64;
foreach $division (4,8,16,32,64,128)
   {
   $place = $portionarray[$arraynumber-1];
   seek(TEXT, $place, 0);
   $line = <TEXT>;
   $line = <TEXT>;
   $line =~ /^([a-z]+) /;
   $estimate_word = $1;
   if ($estimate_word gt $findword)
     {
     $arraynumber = $arraynumber - (128/$division);
     }
   else
     {
     $arraynumber = $arraynumber + (128/$division);
     }
   }
undef ($/);
seek(TEXT,($place - 10000), 0);
read(TEXT,$holder,20000);
if ($holder =~ /\n($findword)[0-9\= ]+\n/)
   {
   print $&;
   }
else
   {
   print "Sorry. Couldn't find $findword in the index.\n";
   }
exit;
```

What is an ordered list? Examples are an index, an alphabetized list such as a dictionary or glossary, or a numbered array (List 6.4.1).

The algorithm for binary searches is very simple. A search item is provided along with the name of a file containing ordered (usually alphabetized) items. Perl opens the file, goes to the middle of the file, and asks itself whether the middle item is higher or lower (in the alphabet) than the search item. If the middle item is lower in the alphabet than the search item, then the search item must reside in the bottom half of the file. Otherwise it resides in the top half of the file. In either case, Perl has reduced the size of the search space by one-half and proceeds to the middle item of the appropriate half-file to repeat the process. In just a few iterations, Perl always manages to get close to the search item.

In the `find_bin.pl` implementation, we start by determining the byte locations of the different binary-power portions of the file. We start searching at the mid-file location. We stop our binary loops after six iterations and locate the search item with a simple regex over a narrowed section of the file.

A typical input and output for the `find_bin.pl` script is shown:

```
C:\ftp>perl find_bin.pl
What word would you like to find?
features
features =  71 76 93 110 114 131 165 259 291
```

Here, the word "features" is our search item. The `find_bin.txt` file consists of alphabetized words followed by lists of numbers (corresponding to page numbers where the words might be found). The script extracts the line containing the word "features" and prints it to the screen.

Binary searches can be performed on any ordered list of any length, at great speed. Because this trick works on files with ordered content, programmers might prefer binary searches on files, rather than database programs, for rapid record look-ups.

6.5. AN ALGORITHM FOR ON-THE-FLY CODED DATA RETRIEVAL WITHOUT PRE-CODING

In this chapter, we have seen how simple it is to find all the occurrences of a phrase or a pattern in a text file or data file. The limitation of this approach comes from the enormous degree of synonymy in biomedical text. Most genes have several different designations. Most diseases have half a dozen equivalent names. If you perform a search on one name, you miss all the occurrences of the same concept appearing under a different name. This is the reason that biomedicine places great emphasis on autocoding (i.e., representing terms with a unique concept number that symbolizes the collection of all terms that have the same meaning).

Let us consider an algorithm that finds all the occurrences of a term, in any text, along with all the occurrences of all the synonyms of a term, without needing to autocode any of the text.

Take your term and go to a nomenclature that is appropriate for your term (i.e., a gene nomenclature for a gene term, a medical nomenclature for the name of a disease, a neoplasm nomenclature for the name of a tumor, etc). In the nomenclature, collect all the term equivalents for your search term. Put every term equivalent in an array. For each member of the array, search the entire data file for all occurrences of the array term. This approach is very simple, but it may have an unacceptable execution time if there are many synonyms for the search term and if the text to be searched is long (Lists 6.5.1 and 6.5.2.).

I previously published an implementation of this algorithm that is very fast (20). It uses any biomedical nomenclature, and any plain-text search file. The algorithm achieves the functionality of a search over an annotated dataset. Because the algorithm requires no pre-coding, it reduces the enormous expenditure of professional time and energy devoted to coding biomedical reports. Speed is attained by implementing the doublet search algorithm (described in Chapter 5). The detailed algorithm and the steps that users must take to implement the algorithm are found in Lists 6.5.3. and 6.5.4.

The Perl scripts and vocabulary files used in the implementation were described in a prior publication (20) and are freely available to the public from the Association for Pathology Informatics (12).These files are also available from the Jones and Bartlett Web site.

LIST 6.5.1. ALGORITHM FOR CODE-BASED SEARCHES WITHOUT PRE-ANNOTATION

The user enters a query term.

All the terms from a preferred nomenclature* that are synonymous with the query term are collected into a list.

Each term in the list is matched against the text corpus to determine the locations in the corpus where the term is found.

A list is assembled of corpus locations matching the query term or its synonyms.

The user's query term is matched against all the equivalent terms included in a preferred vocabulary.

*Any vocabulary is suitable, so long as the vocabulary consists of term/code pairs, where a term and its synonyms are all paired with the same code.

LIST 6.5.2. EXAMPLES OF KEY/VALUES PAIR FOR CODE C0206708

```
C0206708|Cervical Intraepithelial Neoplasms
C0206708|Cervical Intraepithelial Neoplasm
C0206708|Intraepithelial Neoplasm, Cervical
C0206708|Intraepithelial Neoplasms, Cervical
C0206708|Neoplasm, Cervical Intraepithelial
C0206708|Neoplasms, Cervical Intraepithelial
C0206708|Intraepithelial Neoplasia, Cervical
C0206708|Neoplasia, Cervical Intraepithelial
C0206708|Cervical Intraepithelial Neoplasia
```

LIST 6.5.3. IMPLEMENTATION METHOD FOR CODE-BASED SEARCHES WITHOUT PRE-ANNOTATION

1. The doublet index for the dataset is prepared. The doublet index consists of each of the doublets (two consecutive words) in the datasets, along with all the locations in the dataset where each doublet occurs. In the sample dataset used for this manuscript, the doublet term "vancouver canada" occurs 95 times. Examples of 14 index entries for the "vancouver canada" doublet are:

```
vancouver canada = 151198-17
vancouver canada = 157354-8
vancouver canada = 166770-13
vancouver canada = 171565-8
vancouver canada = 175470-11
vancouver canada = 178127-8
vancouver canada = 189527-11
vancouver canada = 198094-8
vancouver canada = 201139-11
vancouver canada = 201398-12
vancouver canada = 202037-8
vancouver canada = 204257-14
vancouver canada = 208131-8
vancouver canada = 223026-11
```

   Each entry consists of the name of the doublet, the record number from the dataset in which the doublet occurs (for example, record number 151198), and the offset position within the record at which the doublet occurs (for example, word number 17). The index can be quickly compiled and, once created, never needs to change unless the dataset changes. This index will be used to locate terms composed of any number of words.

2. An associative array is prepared from the doublet index file, consisting of key/value pairs, where the keys are the set of all the different doublet terms present in the dataset, and the values are the byte locations in the doublet index file where the doublet first occurs. The doublet index file is alphabetized. Knowing the first occurrence of a doublet in the index allows us to quickly find all the index entries for the doublet by simply going to the first location of the doublet and collecting successive line readings from the index file. The collection of all the index entries for the doublet specifies every record and every position in every record where the doublet occurs.
3. The nomenclature of interest is stored in memory as two associative arrays. One associative array consists of key/value pairs, with nomenclature terms as the keys and corresponding nomenclature codes as values. The other associative array consists of key/value pairs with nomenclature codes as the keys and the list of corresponding nomenclature terms as the values (see List 6.5.2). These two associative arrays allow us to quickly match the user's query term against the entire nomenclature, identifying the code number of the matching term (if it exists) and creating an array of all the equivalent terms that match the code.
4. The array of vocabulary terms that are equivalent to the query term entered by the user are consecutively matched against all the records from the dataset (as described in Steps 5, 6, and 7).
5. Each term in the array of equivalent terms is parsed as an array of doublet neighbors. In the case of terms composed of an odd number of words, an overlapping doublet at the end of the term is added. For example, adenocarcinoma of the lung is a term composed of an even number of words.

   The doublet array is: "adenocarcinoma of," "the lung."

   An example of a term composed of an odd number of words is: refractory anemia with excess blasts.

   The doublet array is: "refractory anemia," "with excess," "excess blasts."
6. For each term in the array of equivalent terms, entries are collected from the doublet index if they match any of the consecutive doublets that compose the term. The records that match a term are among the records that match every doublet in the term. For a record in this subset to match the term, it needs to

contain the doublets that compose the term in the same text order as the occurrences of the doublets in the term. The term "refractory anemia with excess blasts" is composed of three doublets that must occur in the following relative word positions: "refractory anemia" ... position n "with excess" ... position n+2 ... "excess blasts" ... position n+3. To determine whether term doublets co-occurring in a record actually match a full term, one needs to test whether the doublets occur in relative positions corresponding to their positions in the term. For instance, a record might contain multiple occurrences of each of the doublets that compose the term "refractory anemia with excess blasts." Imagine that the doublets occur at the following word positions within the record: "refractory anemia" locations (15, 92, 105, 234); "with excess" locations (17, 107, 344); and "excess blasts" locations (18, 108, 992, 1026). We can be certain that the full term occurs twice in the record, beginning in positions 15 and 105. Only at these two offset positions for the doublet "refractory anemia" are there consecutive occurrences in the same record of the ordered doublets in the whole term (that is, offsets n, n+2, n+3).

7. All of the dataset records containing all of the consecutively occurring doublets that compose the terms from the array of terms that are equivalent to the query term are collected from the dataset and annotated with the shared concept vocabulary code.

LIST 6.5.4. USER STEPS FOR ON-THE-FLY CODE SEARCHES

Users will need to have a dataset or plain-text corpus in which every record is separated by the same delimiter. For testing purposes, the author created a 105-megabyte text. The corpus was created by a PubMed query on "pathology (ad)AND neoplasm," at the U.S. National Library of Medicine's Web site (*www.pubmed.org*). The query gathered all abstracts from the PubMed database in which the term neoplasm occurs somewhere in the PubMed entry, and in which the affiliation of the author contains the word "pathology." The query yielded abstracts that are likely to contain names of neoplasms. The PubMed output file can serve as a good test for an autocoder that uses a neoplasm nomenclature. The PubMed search yielded 66,509 abstracts. All of the abstracts were downloaded into a single file from the PubMed site by setting the "Display" attribute to "Abstract" and the "Send to" attribute to "file." This produced a 105,689,546-byte plain-text file. The file was given the filename `tumorab.txt`, and this filename was used by the autocoders as a parsing input file. Although this file is not included with this manuscript, anyone in the world with Internet access can obtain a near-identical file by repeating the same PubMed query. The records from the text consisted of paragraphs delimited by double return characters (also referred to as double-newline characters or double ASCII(13)-ASCII(10) character pairs). This is a common way of delimiting textual paragraphs.

Users will create a file of all the doublets in the file, one line to each doublet, and each doublet followed by the record number and the record word-offset for the doublet occurrence. When a doublet occurs several times in a single record, each occurrence is indexed, with a different word-offset for each occurrence. The Perl script that creates the doublet index is `doubdat2.pl`. It might require several minutes to execute, and it produces an index file, that is 400+ megabytes in length. This file is, in turn, sorted alphabetically by `bigsort.pl`, a short Perl script that sorts files of any size. The index file is named `doubdat.out`. `Doubdat2.pl` also keeps track of the begin-byte location in `tumorab.txt` where each text record begins. Another Perl script, `doub4.pl`, creates a database file containing the byte location in `doubdat.out` for the first occurrences of doublets. `Annotget.pl` performs term searches on 100 terms selected randomly from a chosen vocabulary. In this case, the vocabularies tested are Snomed-CT, extracted from the 2004 version of UMLS, (21), (22), (23), (24), and the "Taxonomy for the Developmental Lineage Classification of Neoplasms," made available for public download by the Association for Pathology Informatics (12).

CHAPTER

7

Cryptography and De-identification

7.1. BACKGROUND

The field of cryptography encompasses a variety of useful techniques (List 7.1.1) (25), (26). Most of these methods involve using a one-way hash algorithm or a pseudo-random number generator. A one-way hash algorithm can be easily implemented in Perl programs using either the MD_5 or the SHA modules. A reasonably good random number generator is available with Perl's built-in `rand()` operator. We will be using random number generators extensively in Chapter 14.

The purpose of this chapter is to explain how readily available cryptographic techniques can be used in Perl scripts to provide powerful and elegant solutions to biomedical problems related to identification, de-identification, data authentication, and data privatization.

LIST 7.1.1. CRYPTOGRAPHIC METHODS VITAL TO BIOMEDICINE

Encrypting and decrypting messages.
Electronic signatures.
Message authentication.
Time stamping.
Creating unique identifiers.
Reconciling patients across institutions.
De-identification and Re-identification.
Privatizing data sharing protocols.
Data referencing (with message digests).
Watermarking and steganography utilities.

7.2. PERL CRYPTOGRAPHY MODULES

LIST 7.2.1. PERL CRYPTOGRAPHY MODULES

Authen-DigestMD5
Crypt-Caesar
Crypt-Chimera
Crypt-Cipher
Crypt-Discordian
Crypt-Enigma
Crypt-GeneratePassword
Crypt-PasswdMD5
Crypt-RandPasswd
Crypt-Rot13
Crypt-Salt
Crypt-Solitaire
Crypt-Vigenere
Digest-MD5-File
Digest-MD5-Reverse
Digest-Perl-MD4
Digest-Perl-MD5
Digest-SHA
Digest-SHA-PurePerl
Email-Auth-AddressHash
SHA

7.3. ONE-WAY HASHING ALGORITHMS

A one-way hash is an algorithm that transforms a string into another string is such a way that the original string cannot be calculated by operations on the hash value (hence the term "one-way" hash). A string will always yield the same one-way hash value whenever the hash algorithm is applied. One-way hash algorithms are sometimes called HMACs (Hashed Message Authentication Codes) or message digests, and they should not be confused with Perl hashes (associative arrays). Examples of public domain one-way hash algorithms are MD5 (28) and SHA, the Secure Hash Algorithm (29). These differ from encryption protocols that produce an output that can be decrypted by a second computation on the encrypted string.

The resultant one-way hash values for text strings consist of near-random strings of characters, and the length of the strings (e.g., the strength of the one-way hash) can be made arbitrarily long.

One-way hashes can be used to anonymize patient records while still permitting researchers to accrue data over time to a specific patient's record (List 7.3.1.).

LIST 7.3.1. EXAMPLE PROTOCOL FOR A ONE-WAY HASH, DE-IDENTIFIED RECORD LINKAGE

1. John Q. Public arrives for the first time in your medical clinic.
2. John Q. Public has a glucose test ordered and receives a glucose value of 85.
3. Using the MD_5 one-way hash algorithm, on the character string, "John Q. Public," a hash value of "3f875ec450dfbb07ed889e7b9c36da92" is generated.
4. In addition to John Q. Public's identified medical record, a de-identified record is prepared:
   3f875ec450dfbb07ed889e7b9c36da92^^glucose^^85

   A property of the one-way hash value is that it is a seemingly random collection of letters and numbers, and no computational efforts applied to the one-way hash value can yield the patient's name.

   The de-identified record is given to a trusted database administrator who adds it to the database of de-identified records. The database administrator cannot identify any of the patients whose records are included in the database.
5. Ten years later, John Q. Public returns to the medical clinic and has another glucose test. This time, the glucose value is 95.

   A one-way hash is performed on the string "John Q. Public" yielding 3f875ec450dfbb07ed889e7b9c36da92, and a new de-identified record is prepared:
   3f875ec450dfbb07ed889e7b9c36da92^^glucose^^95

   The de-identified record is given to the trusted database administrator, who adds it to the aggregate database. The database program finds a match to the one-way hash and concatenates the new record to the old record:

   3f875ec450dfbb07ed889e7b9c36da92^^glucose^^85^^glucose^^95

What has this accomplished? It achieves the seemingly impossible feat of accruing clinical data over time for de-identified data records.

7.4. ONE-WAY HASH WEAKNESSES: DICTIONARY ATTACKS AND COLLISIONS

Insightful readers will notice that this approach has a flaw. Attacks on one-way hash data might take the form of hashing a list of names and looking for matching hash values in the dataset (a so-called dictionary attack). Efforts to overcome this limitation include encrypting the hash, hashing a secret combination of identifier elements, or both, or keeping the hash value private (hidden). As in any privacy protocol, success is achieved when implementation strategies minimize risk through a realistic assessment of local security weaknesses. Regarding implementation, problems arise when institutions have a flawed system for identifying patients. If a person is identified within a hospital system as Tom Peterson on Monday, and Thomas Peterson on Tuesday, a one-way hash strategy based on patient names would certainly fail. If the hash is performed on unique, persistent patient identifiers, such an approach would have a better chance of success.

Technical problems might also arise. One-way hash collisions occur when two different strings yield the same hash value. Because hash values are pseudo-random character strings, the chance of a hash collision between two patients with different identifiers is very small. A variety of solutions has been suggested for large database implementations (where collisions might rarely occur). The most straightforward maneuver is to use a longer hash value. SHA (Secure Hash Algorithm) has different algorithmic forms (SHA-1, SHA-256, SHA-384, and SHA-512) with message digest (hash) lengths up to 512 bits. As the length of the message digest increases, the chance of having a digest collision becomes very small.

A 2005 paper by Faldum and Pommerening proposes a novel approach to assigning patient identifiers (27). They propose a mathematical algorithm that can distinguish one billion individuals without collisions, using a short, eight-character string that contains extra characters for error checking and correcting.

7.5. COMPUTING A ONE-WAY HASH FOR A WORD, PHRASE, OR FILE

Perl is distributed with modules that will create one-way hashes for any character strings and for whole files, as shown (List 7.5.1).

This `md5.pl` script invokes the Perl MD5 module:

```
use MD5;
```

LIST 7.5.1. PERL SCRIPT, `md5.pl` CREATES AN MD5 ONE-WAY HASH VALUE FOR ANY PROVIDED STRING

```
#!/usr/local/bin/perl
use MD5;
print "What words would you like to digest?\n";
$holdstring = <STDIN>;
chomp;
$hexhashstring = MD5->hexhash($holdstring);
print "md_5 hexhash => $hexhashstring\n";
exit;
```

This command line instructs Perl to use the MD5 one-way hash module that is bundled in standard Perl distributions.

The next line prompts the user to provide a character string. The user enters any string, and the script uses the chomp command to cut off the newline character at the end of the input string.

The script next calls the MD5 module's hexhash method:

```
$hexhashstring = MD5->hexhash($holdstring);
```

As usual, we looked inside the module and found the usage statement.

The MD5 module file is bundled in the standard Perl distribution. A usage statement in the file indicates how to call the hexhash subroutine to transform a scalar into a one-way hashed output string in hex notation:

```
$string = MD5->hexhash(SCALAR);
```

Hex notation is base-16 ASCII, and uses an alphabet composed of: 0,1,2,3,4,5,6,7,8,9,a,b,c,d,e,f.

The output of the MD5 algorithm for two rounds of input using the same character string, "Jules Berman" is shown (List 7.5.2).

The first two executions demonstrate that when an input string is hashed, it produces the same output with each execution of the script. The third execution shows that if any characters of an input string are changed, it produces a completely different one-way hash output.

One-way hashes are useful for authenticating files. A file can be hashed, just as a character string can be hashed, producing a string of characters that is virtually unique for the file. If a single byte in the file is changed, the one-way hash for the file is completely changed. If a file always produces the same one-way hash value, it's a safe bet that the file has not been modified. The Perl MD5 module has a method for producing one-way hashes on files (List 7.5.3).

LIST 7.5.2. THREE EXECUTIONS OF THE MD_5 ALGORITHM

```
Execution 1:
c:\ftp>perl md5_word.pl
What words would you like to digest?
Jules Berman
md_5 hexhash => 0ab7ad79962fd2ea036cc8dbaade6f2a


Execution 2:
c:\ftp>perl md5_word.pl
What words would you like to digest?
Jules Berman
md_5 hexhash => 0ab7ad79962fd2ea036cc8dbaade6f2a


Execution 3:
c:\ftp>perl md5_word.pl
What words would you like to digest?
Jules J. Berman
md_5 hexhash => b59d141b7962b930e7a803bbaa451ddf
```

LIST 7.5.3. CREATING AN MD_5 ONE-WAY HASH FOR A FILE

```
#!/usr/local/bin/perl
use MD5;
print "What file would you like to digest?\n";
$holdfile = <STDIN>;
chomp;
open (TEXT,"$holdfile");
$context = new MD5;
$context->addfile(TEXT);
$digest = $context->digest();
print (unpack ("H*", $digest));
exit;
```

7.6. THRESHOLD PROTOCOL

Threshold cryptographic protocols divide messages into multiple pieces, with no single piece containing information that can reconstruct the original message (25). Threshold protocols have been used since antiquity, commonly appearing as plot devices in adventure novels. A map to buried treasure is divided among the central characters; a puzzle is reconstructed when five missing pieces are assembled; measured turns of the combination lock are distributed among untrustworthy co-conspirators; matching rings in a set are destroyed. We will review a simple threshold protocol that can be used to search, annotate, or transform confidential data without breaching patient confidentiality (List 7.6.1)(30).

LIST 7.6.1. THE BASIC THRESHOLD PROTOCOL

1. Text is divided into short phrases separated by high frequency words (words that occur at high frequency in any text).
2. Each phrase is converted by a one-way hash algorithm into a seemingly random set of characters.
3. Threshold Piece 1 is composed of the list of all phrases, with each phrase followed by its one-way hash.
4. Threshold Piece 2 is composed of the text, with all phrases replaced by their one-way hash values, and with high-frequency words preserved.

The threshold protocol yields two threshold pieces with the following properties:

Neither Piece 1 nor Piece 2 contains confidential information.

The original text can be reconstructed from Piece 1 and Piece 2.

Let us take a step back to see how the threshold protocol might actually apply. A generalized confidentiality problem can be presented as a negotiation protocol between Alice and Bob. Bob has a file containing the medical records of millions of patients. Alice has secret software that can annotate Bob's file, enhancing its value manyfold. Alice won't give Bob her secret algorithm, but is willing to demonstrate the algorithm if Bob gives her his database. Bob won't give Alice the database, but he can give her little snippets of the database containing insufficient information to infer patient identities.

Bob prepares an algorithm that transforms his file into two threshold pieces. Piece 1 is a file that contains all of the phrases from the original file with each phrase attached to its one-way hash value. A one-way hash value is a character string composed of a fixed number of seemingly random characters selected by a mathematical algorithm that cannot be reversed (25).

As discussed in the prior section, a one-way hash has two important properties:

1) A phrase will always yield the same hash value when the one-way hash algorithm operates on the phrase.
2) There is no feasible way to determine the phrase by inspecting or manipulating the hash value. This second property holds true even if the hashing algorithm is known.

Bob will give Alice Piece 1.

Piece 2 is a file wherein each phrase from the original file is replaced by its one-way hash value. High frequency words (so-called "stop" words such as the, and, an, but, if, etc.) are left in place in Piece 2. The use of "stop" words to extract useful phrases from text is a popular indexing technique. The list of "stop" words used in the threshold algorithm was taken directly from the National Library of Medicine's PubMed resource.

> *http://www.ncbi.nlm.nih.gov/books/bv.fcgi?rid=helppubmed.table.pubmedhelp.T42*

Piece 2 will be used to reconstruct the original text or an annotated version of the original text, using Alice's modifications to Piece 1.

The following is an example of a single line of Bob's text that has been converted into two threshold pieces according to the described algorithm.

Bob's original text:

> "They suggested that the manifestations were as severe in the mother as in the sons and that this suggested autosomal dominant inheritance."

If Alice had Piece 1 and Piece 2, she could simply use Piece 1 to find the text phrases that match the hash-values in Piece 2. Substituting the

LIST 7.6.2. BOB'S PIECE 1

```
684327ec3b2f020aa3099edb177d3794 => suggested autosomal
   dominant inheritance
3c188dace2e7977fd6333e4d8010e181 => mother
8c81b4aaf9c2009666d532da3b19d5f8 => manifestations
db277da2e82a4cb7e9b37c8b0c7f66f0 => suggested
e183376eb9cc9a301952c05b5e4e84e3 => sons
22cf107be97ab08b33a62db68b4a390d => severe
```

LIST 7.6.3. BOB'S PIECE 2

```
they db277da2e82a4cb7e9b37c8b0c7f66f0 that the
8c81b4aaf9c2009666d532da3b19d5f8 were as
22cf107be97ab08b33a62db68b4a390d in the
3c188dace2e7977fd6333e4d8010e181 as in the
e183376eb9cc9a301952c05b5e4e84e3 and that this
684327ec3b2f020aa3099edb177d3794.
```

phrases back into Piece 2 will re-create Bob's original line of text. Bob must ensure that Alice never obtains Piece 2.

The negotiation between Alice and Bob:

Bob prepares threshold Pieces 1 and 2 and sends Piece 1 to Alice. Alice might require Bob to prove the authenticity of Piece 1, but Bob has no reason to care if an unauthorized party intercepts Piece 1. Alice uses her software (which might be secret, or it might require computational facilities that Bob doesn't have, or it might require large databases that Bob doesn't have), to transform or annotate each phrase from Piece 1. The transformation product for each phrase can be almost anything that

LIST 7.6.4. PROPERTIES OF PIECE 1 (THE LISTING OF PHRASES AND THEIR ONE-WAY HASHES)

Contains no information on the frequency of occurrence of the phrases found in the original text (because recurring phrases map to the same hash code and appear as a single entry in Piece 1).

Contains no information that Alice can use to connect any patient to any particular patient record. Records do not exist as entities in Piece 1.

Contains no information on the order or locations of the phrases found in the original text.

Contains all the concepts found in the original text. Stop words are a popular method of parsing text into concepts.

Bob can destroy Piece 1 and re-create it later from the original file.

Alice can use the phrases in Piece 1 to transform, annotate, or search the concepts found in the original file.

Alice can transfer Piece 1 to a third party without violating HIPAA privacy rules or Common Rule human subject regulations (in the United States). For that matter, Alice can keep Piece 1 and add it to her database of Piece 1 files collected from all of her clients.

LIST 7.6.5. PROPERTIES OF PIECE 2

Contains no information that can be used to connect any patient to any particular patient record.

Contains nothing but hash values of phrases and stop words, in their correct order of occurrence in the original text.

Anyone obtaining Piece 1 and Piece 2 can reconstruct the original text.

Bob can lose or destroy Piece 2, and re-create it later from the original file.

Bob considers valuable (for example, a UMLS code, a genome database link, an image file URL, or a tissue sample location). Alice substitutes the transformed text (or simply appends the transformed text) for each phrase back into Piece 1, collocating it with the original one-way hash value associated with the phrases.

Let us pretend that Alice has an autocoder that provides a standard nomenclature code to medical phrases that occur in text. Alice's software transforms the original phrases from Piece 1, preserving the original hash values. Phrases from Piece 1 that occur in the Unified Medical Language System now have been given code numbers by Alice's software:

684327ec3b2f020aa3099edb177d3794 => suggested (autosomal dominant inheritance=C0443147)

3c188dace2e7977fd6333e4d8010e181 => (mother=C0026591)

8c81b4aaf9c2009666d532da3b19d5f8 => manifestations

db277da2e82a4cb7e9b37c8b0c7f66f0 => suggested

e183376eb9cc9a301952c05b5e4e84e3 => (son=C0037683)

22cf107be97ab08b33a62db68b4a390d => (severe=C0205082)

Alice returns the coded phrase list (above) from Piece 1 to Bob. Bob now takes the transformed Piece 1 and substitutes the transformed phrases for each occurrence of the hash values occurring in Piece 2 (which he has saved for this very purpose).

The reconstructed sentence is now:

they suggested that the manifestations were as (severe=C0205082) in the (mother=C0026591) as in the (son=C0037683) and that this suggested (autosomal dominant heritance=C0443147).

The original sentence is now annotated with UMLS codes. It was accomplished without sharing confidential information that might have been contained in the text. Bob never had access to Alice's software. Alice never had the opportunity to see Bob's original text. A Perl implementation of the algorithm is shown in List 7.6.6. The script requires any external plain-text file, named AA.TXT in the script, as the source of the threshold pieces.

LIST 7.6.6. PERL SCRIPT, thresh.pl, IMPLEMENTING THE THRESHOLD ALGORITHM

```
#!/usr/bin/perl
use MD5;
$start = time();
my %stoparray = qw(
     a 1 about 1 again 1 all 1 almost 1 also 1 although 1 always 1
     among 1 an 1 and 1 another 1 any 1 are 1 as 1 at 1 be 1
     because 1 been 1 before 1 being 1 between 1 both 1 but 1 by
     1 can 1 could 1 did 1 do 1 does 1 done 1 due 1 during 1
     each 1 either 1 enough 1 especially 1 etc 1 for 1 found 1
     from 1 further 1 had 1 has 1 have 1 having 1 here 1 how 1
     however 1 i 1 if 1 in 1 into 1 is 1 it 1 its 1 itself 1
     just 1 kg 1 km 1 made 1 mainly 1 make 1 may 1 mg 1 might 1
     ml 1 mm 1 most 1 mostly 1 must 1 nearly 1 neither 1 no 1
     nor 1 obtained 1 of 1 often 1 on 1 our 1 overall 1 perhaps 1
     quite 1 rather 1 really 1 regarding 1 seem 1 seen 1 several 1
     should 1 show 1 showed 1 shown 1 shows 1 significantly 1
     since 1 so 1 some 1 such 1 than 1 that 1 the 1 their 1
     theirs 1 them 1 then 1 there 1 therefore 1 these 1 they 1
     this 1 those 1 through 1 thus 1 to 1 upon 1 use 1 used 1
     using 1 various 1 very 1 was 1 we 1 were 1 what 1 when 1
     which 1 while 1 with 1 within 1 without 1 would 1 or 1
     can't 1 doesn't 1 not 1
);
open(TEXT, "AA.TXT");
open (ONEOUT, ">AA1.TXT");
open (TWOOUT, ">AA2.TXT");
$k = " ";
while ($k ne "")
  {
  $phrase = "";
  $k = <TEXT>;
  $y = $k;
```

```
$y = lc($y);
$y =~ s/\.\n//;
@karray = split(/ +/, $y);
foreach my $value (@karray)
    {
    if (exists $stoparray{$value})
      {
      $phrase =~ s/^ +([^ ])/$1/;
      $phrase =~ s/ +/ /g;
      $phrase =~ s/ +$//;
      if ($phrase ne "")
        {
        $hashstring = MD5->hash($phrase);
        $hexhashstring = MD5->hexhash($hashstring);
        $index{$hexhashstring} = $phrase;
        print TWOOUT "$hexhashstring $value ";
        $phrase = "";
        }
      else
        {
        print TWOOUT " $value ";
        }
      }
    else
      {
      $phrase = "$phrase $value ";
      }
    }
if ($phrase ne "")
  {
  $phrase =~ s/^ +([^ ])/$1/;
  $phrase =~ s/ +/ /g;
  $phrase =~ s/ +$//;
  if ($phrase ne "")
    {
    $hashstring = MD5->hash($phrase);
    $hexhashstring = MD5->hexhash($hashstring);
    $index{$hexhashstring} = $phrase;
    print TWOOUT "$hexhashstring\.\n";
    }
  }
if ($phrase eq "")
  {
```

```
     print TWOOUT "\n";
     }
   $phrase = "";
   }
while ((my $key, my $value) = each(%index))
     {
     print ONEOUT "$key -> $value\n";
     }
$total = time() - $start;
print "\nThe total time was $total seconds.";
close TEXT;
close ONEOUT;
close TWOOUT;
open (TWOOUT, "AA2.TXT");
open (OUTOUT, ">AA3.TXT");
$line = " ";
while ($line ne "")
   {
   $line = <TWOOUT>;
   $fixline = $line;
   $fixline =~ s/\.\n//;
   @linearray = split(/ +/,$fixline);
   foreach my $value (@linearray)
     {
     if (exists $index{$value})
       {
       print OUTOUT "$index{$value} ";
       }
     else
       {
       print OUTOUT "$value ";
       }
     }
   }
exit;
```

7.7. IMPLEMENTATION ISSUES FOR THE THRESHOLD PROTOCOL

Depending on the type of file that needs to be converted into threshold pieces, some data preparation might be useful. In particular, it might be useful to encrypt or delete specific identifiers found in the original file, such as surgical pathology numbers. The file that is used by the algorithm should itself be assigned a hash number by the algorithm, as should file

1 and file 2. These three hash numbers could be saved and used for authentication purposes in later stages of a data-negotiation protocol.

The original text is converted into two pieces, neither of which contains any identifying information. There is sufficient information in Piece 1 for Alice to annotate the text and return it to Bob (annotated Piece 1). Bob can reconstruct his original text, including Alice's annotations, thus adding value to his original data, without breaching patient confidentiality. Bob can pay Alice for her services. Alice can keep Piece 1 and use it for her own purposes. Alice can make a large database consisting of all the Piece 1 files she receives from all of her customers. Alice can sell Piece 1 to a third party, if she wants. Alice can update or otherwise enhance her annotations on Piece 1 and sell the updated files to Bob.

The same protocol could have been implemented in a three-party negotiation. Bob might have been a data supplier with no interest in using the data himself. Suppose Carol was interested in Alice's annotations of Bob's file. Bob might have given Alice threshold Piece 1 and Carol threshold Piece 2. Alice might have made her transformation of the phrases in Piece 1 and sent the transformed version of Piece 1 to Carol. Carol could use Alice's transformed version of Piece 1 and her copy of Piece 2 to create a transformed version of Bob's original text. This would only work, of course, if the transformed version of Bob's original file (produced by Carol) contains no confidential information. A variation might involve assigning Bob as the trusted broker, who uses Piece 2 and the transformed version of Piece 1 to create a file for Carol. In this variation, Carol receives nothing until the end of the negotiation and Bob can take measures to ensure that the file that Carol receives is "safe."

The threshold negotiation need not be based on text exchange. The same negotiation would apply to any set of data elements that can be transformed or annotated. The threshold protocol has greatest practical value in instances when data elements inform on other data elements that reside in the same data record. The protocol teases apart the data records and substitutes one-way hash values back into the record. The ways in which individual pieces of data can be transformed or annotated are limited only by the imagination. As an example, sequences of DNA can be annotated with positional mappings, standard nomenclature codes, or similarity information.

CHAPTER

8

Scrubbing Data

8.1. BACKGROUND

One of the biggest challenges in biomedicine today is data privacy and security. Because medical data is often extracted from private, legally protected medical records, people who work with medical data must be versed in the legal and ethical issues related to medical data protections. In the United States, private medical data is protected by two federal regulations: HIPAA and the Common Rule ((31), (32)).

HIPAA and the Common Rule both permit sharing of confidential medical data when the patient is informed of the research risks and provides consent. When a research project uses a small number of medical records, it might be feasible to obtain patient consent for each record. However, if large numbers of records are needed, the consent option is impractical. In the absence of patient consent, HIPAA and the Common Rule permit research on pre-existing records that have been stripped of all information that could identify the patient.

Latanya Sweeney was an early proponent of technical approaches to medical record de-identification and has published extensively on the subject (33), (34), (35). Her work formed the foundation for current multistep approaches to de-identification encompassing the following tasks:

LIST 8.1.1. MAKING MEDICAL RECORD DATA HARMLESS

1. De-identification of data fields that specifically characterize the patient (name, Social Security number, hospital number, address, age, etc.)
2. Free-text data scrubbing, removing identifiers from the textual portion of medical reports (36)

3. Rendering the dataset ambiguous, ensuring that patients cannot be identified by data records containing a unique set of characterizing information
4. Free-text data privatizing, removing any information of a private nature that might be contained within the report

HIPAA lists 18 identifiers that must be absent from any de-identified medical records (List 8.1.2).

Data scrubbing involves removing HIPAA identifiers and private information that is present in medical records (36). What is "private" text? Private text is text that is nobody's business and that does not enhance the intended use of the de-identified patient record. In many cases, private text is written by hospital personnel with the expectation that it will be shared only among the persons directly responsible for the care of the patient. This might include notes documenting errors, misjudgments, warnings, and complaints. Most hospital personnel are

LIST 8.1.2. HIPAA-SPECIFIED IDENTIFIERS

- Names
- Geographic subdivisions smaller than a state
- Dates (except year) directly related to patient
- Telephone numbers
- Fax numbers
- E-mail addresses
- Social Security numbers
- Medical record numbers
- Health plan beneficiary numbers
- Account numbers
- Certificate/license numbers
- Vehicle identifiers and serial numbers
- Device identifiers and serial numbers
- Web URLs
- Internet Protocol (IP) address numbers
- Biometric identifiers, including finger and voiceprints
- Full-face photographic images and any comparable images
- Any other unique identifying number, characteristic, or code, except as permitted under HIPAA to re-identify data

expected to exclude information of an incriminating nature from the patient's medical records. Incident reports and quality assurance reports exist for this purpose. In reality, medical records often contain information that is best removed from shared data sets. An exception list of offending terms that must be removed, when present in text, is a first-pass remedy.

There is a general understanding that when medical data are shared for the purposes of conducting research, there is an ethical obligation to share only that portion of the patient record that is actually needed to conduct the research. This is sometimes referred to as the "minimum necessary" principle (see Glossary).

One method of data scrubbing is the subtractive method. In this method, the text is parsed, sentence-by-sentence, and the computer program extracts words and numbers that match any item from an exclusion list (List 8.1.3).

Subtractive scrubbers typically filter text through dozens or even hundreds of regex expressions, extracting text that matches patterns consistent with dates, times, words followed by honorifics (for example, Mr., Ms., Dr.), and certain data formatted within the report in a manner specific to the institution (for example, report accession numbers).

Subtractive methods mimic the way that a human might censor a text. The human reads the text with a marker in hand, prepared to strike out any words or phrases that she knows is offensive. In my opinion, subtractive methods for data scrubbing are ineffective and counterproductive (List 8.1.4).

Two other approaches to data scrubbing are the concept-match method and the doublet-preserving method, both of which are fast and do not rely on exclusion lists.

The purpose of this chapter is to provide simple, effective Perl techniques to scrub and de-identify confidential biomedical data.

LIST 8.1.3. SOME ITEMS THAT MIGHT BE KEPT IN EXCLUSION LISTS

- List of patient names
- List of staff names
- List of patient hospital identifiers or Social Security numbers
- List of hospitals, departments, wards, and rooms
- Geographic locations, including street names, and zip codes
- List of expletives

LIST 8.1.4. DEFICIENCIES OF SUBTRACTIVE DATA-SCRUBBING METHODS

- Require the creation and continuous maintenance of an identifier list consisting of names of patients, staff, and medical centers as well as addresses and other geographic minutiae.
- Require the creation and continuous maintenance of rules for excluding text based on co-locations or patterns of expression that might signify a HIPAA identifier (for example, a sequence of digits and slashes that might represent a date).
- Do not exclude private information that is non-identifying, but which might be incriminating or distasteful.
- Do not satisfy the "minimum necessary" (see Glossary) principle, holding that medical data convey only that information that is needed for research purposes.
- It is slow. Each parsed sentence is typically evaluated through the entire list of pattern rules. This means that parsing a long corpus of medical text will take considerable time.
- It is complex. Maintaining the rule list and the identifier list will add to the overall complexity of the software. Each institution that implements the software will need to maintain their own lists created for their patients and for their textual styles and formats.
- It is inadequate. Subtractive scrubbers, under the best of circumstances, will occasionally miss an identifier. If a scrubber is 99 percent accurate, it might miss thousands of identifiers in a large text.

8.2. SCRUBBING TEXT USING THE CONCEPT-MATCH METHOD

The algorithm for the Concept-Match scrubber is simple to list, but each of the steps is a complex computational task (37).

1. Parse all input into sentences.
2. Parse each sentence into words.
3. Each stop word (high-frequency words, including prepositions and common adjectives) is preserved in its original place within each sentence.
4. Intervening words and phrases are mapped to a standard nomenclature. This step requires breaking phrases into all possible ordered concatenations of words. For instance, "Margins free of tumor" would become "margins free of tumor, margins free of, free of tumor, margins free, free of, of tumor, margins, free, of, tumor."

Each member of the derivative list is matched against a nomenclature, such as the Unified Medical Language System (UMLS), to determine if it matches any of the terms in the nomenclature. Large terms subsume smaller substring terms. For instance, an occurrence of "adenocarcinoma of prostate" in a sample text may match against "adenocarcinoma", "prostate" and "adenocarcinoma of prostate" in a standard nomenclature. Only the largest matching term, "adenocarcinoma of prostate" will be used.

5. Each coded term is replaced by a synonymous term from the nomenclature, if a synonym exists. For instance, the term renal cell carcinoma appearing in the text might be replaced by the equivalent term, hypernephroma. This step produces an output containing a set of words different from the original text.
6. All other words are replaced by blocking symbol (consisting of three asterisks).

The algorithm is nomenclature-independent. Any standard medical terminology that provides codes for medical concepts and synonyms for all the terms belonging to a medical concept can be used for the purposes of data scrubbing.

LIST 8.2.1. PERL SNIPPET FROM `abscrub.pl` CREATES HASHES FOR STOP WORDS AND NOMENCLATURE TERMS

```
     #!/usr/local/bin/perl
     my %stoparray = qw(
a 1 about 1 absent 1 absence 1 again 1 all 1 almost 1 also 1
although 1 always 1 among 1 an 1 and 1 another 1 any 1 are 1 as 1
at 1 be 1 because 1 been 1 before 1 being 1 between 1 both 1 but 1
by 1 can 1 could 1 cm 1 did 1 do 1 does 1 done 1 due 1 during 1
each 1 either 1 enough 1 especially 1 etc 1 for 1 found 1 from 1
further 1 had 1 has 1 have 1 having 1 here 1 how 1 however 1 i 1 if
1 in 1 into 1 is 1 it 1 its 1 itself 1 just 1 kg 1 km 1 made 1
mainly 1 make 1 may 1 mg 1 might 1 ml 1 mm 1 most 1 mostly 1 must 1
nearly 1 neither 1 no 1 nor 1 observed 1 obtained 1 of 1 often 1 on
1 our 1 overall 1 perhaps 1 present 1 presence 1 quite 1 rather 1
really 1 regarding 1 seem 1 seen 1 several 1 should 1 show 1 showed
1 shown 1 shows 1 significantly 1 since 1 so 1 some 1 such 1 than 1
that 1 the 1 their 1 theirs 1 them 1 then 1 there 1 therefore 1
these 1 they 1 this 1 those 1 through 1 thus 1 to 1 upon 1 use 1
used 1 using 1 various 1 very 1 was 1 we 1 were 1 what 1 when 1
which 1 while 1 with 1 within 1 without 1 would 1 or 1 can't 1
doesn't 1 not 1 );
#This is an associative array containing a list of so-called stop words
#that occur in high frequency in every text
```

```
open (TEXT,"neocl.xml")||die"Cannot";
#neocl.xml is a free, open access cancer nomenclature containing
#about 145,000 different names of tumors arranged by synonymy and
#by tumor class.
my $line = " ";
my $count = 1;
my (%literalhash, %doubhash);
while ($line ne "")
  {
  my ($doublet, $phrase, $code, $i);
  $line = <TEXT>;
  $line =~ /\"(C[0-9]{7})\"/;
  $code = $1;
  $line =~ /\"\> ?(.+) ?\<\//;
  #The preceding lines of code simply extract the terms
  #of the neocl.xml nomenclature.  See Appendix.
  #The following 7 lines remove some unnecessary punctuation,
  #plural forms of tumors and extra space characters
  $phrase = $1;
  $phrase =~ s/\'s//g;
  $phrase =~ s/\,/ /g;
  $phrase =~ s/[^a-z0-9 \-]/ /g;
  $phrase =~ s/\b([a-z]+oma)s/$1/g;
  $phrase =~ s/\b(tumo[u]?r)s/$1/g;
  $phrase =~ s/ +/ /g;
  if ($phrase !~ / /)
     {
     $literalhash{$phrase} = $code;
     next;
     }
  #The terms from the nomenclature are placed into an
  #associative array, as keys, and the codes corresponding
  #to the terms are assigned as values for the keys
```

Because the Perl script is somewhat long (over 9,000 bytes) only a snippet of the `abscrub.pl` script is shown. The entire script can be downloaded from the Jones and Bartlett book Web site. The provided snippet (List 8.2.1.) shows how the associative arrays for stop words and for nomenclature terms are built. The remainder of the script (not shown) simply parses through the text, preserving the words and concepts in the text that match stop words or nomenclature terms, and replacing everything else with asterisks. The resulting text is de-identified and scrubbed.

8.3. COMPOSING A LARGE CORPUS OF SCRUBBED AND CODED MEDICAL TEXT

The Perl script, abscrub.pl, described in the prior section, requires a corpus of text to be scrubbed. In Chapter 3, we showed how a PubMed query could provide text from which email addresses could be extracted. A Pubmed query on any desired key words can produce a large corpus of biomedical text containing abstracts, the summary paragraph of published scientific articles. To test the Perl scrubber, we used a large collection of abstracts extracted from a PubMed search (abstract.txt, see List 8.3.1).

LIST 8.3.1. FILES USED IN abscrub.pl SCRIPT

```
ABSTRACT.TXT    15,232,231 bytes (corpus of PubMed abstracts)
ABS_1.XML       20,190,120 bytes (output file, see List 8.3.3.)
NEOCL.XML        9,867,320 bytes (see Appendix for file details)
```

LIST 8.3.2. FUNCTIONS OF THE abscrub.pl SCRIPT

- Scrubs the 15+-megabyte file of abstracts downloaded from PubMed, abstract.txt
- Produces an output file, abs_1.xml
- High frequency words (such as a, the, of, if) are retained in the output file. Terms from the file that exactly match the terms from an external nomenclature are retained. The script can be adopted to retain terms from any vocabulary or from combined sets of different nomenclatures. All other words from the scrubbed text are replaced by "*". The script simultaneously autocodes the terms from the abstracts while it scrubs the file (see Chapter 5).

The output of the abscrub.pl script is a scrubbed and autocoded transformation of the original text, in XML format.

No identifiers will be present in the output because the only words in the output come from an approved list of terms. But the output of the algorithm can be hard to read. An asterisk will replace identifying words in the original text. The text might consist predominantly of asterisks if it contains many words and terms that are not present in the "approved" word list.

LIST 8.3.3. SAMPLE OUTPUT FROM `abscrub.pl`

```
<?xml version="1.0"?>
<rdf:RDF
     xmlns:rdf="http://www.w3.org/1999/02/22-rdf-syntax-ns#"
     xmlns:dc="http://www.purl.org/dc/elements/1.0/"
xmlns:v="http://www.pathologyinformatics.org/informatics_r.htm">
    <rdf:Description about="urn:PMID-16160487">
        <dc:title>
interobserver and intraobserver variability in the diagnosis
         of hydatidiform mole
        </dc:title>
        <v:autocode term="mole" code="C0000000" />
        <v:autocode term="hydatidiform mole" code="" />
<de_id> * * and * * in the * of hydatidiform mole **</de_id>
    </rdf:Description>
    <rdf:Description about="urn:PMID-16160486">
        <dc:title>
        primary glial tumor of the retina with features of
        myxopapillary ependymoma
        </dc:title>
        <v:autocode term="tumor" code="C0000000" />
<v:autocode term="myxopapillary ependymoma" code="C0000000"/>
<v:autocode term="tumor of the retina" code="C0000000" />
        <v:autocode term="glial tumor" code="C3059000" />
        <v:autocode term="ependymoma" code="C0000000" />
<de_id> * * glial tumor of the retina with * of
   myxopapillary ependymoma * *</de_id>
    </rdf:Description>
    <rdf:Description about="urn:PMID-16160485">
        <dc:title>
cd20-negative t-cell-rich b-cell lymphoma as a progression
of a nodular lymphocyte-predominant Hodgkin's lymphoma treated
with rituximab a molecular analysis using laser capture
        microdissection
        </dc:title>
        <v:autocode term="lymphoma" code="C0000000" />
        <v:autocode term="Hodgkin's" code="C0000000" />
        <v:autocode term="b-cell lymphoma" code="C6858100" />
<v:autocode term="t-cell-rich b-cell lymphoma"
   code="C9496100" />
        <v:autocode term="Hodgkin's lymphoma" code="" />
```

```
<de_id> * * t-cell-rich b-cell lymphoma as a * of a * *
   Hodgkin's lymphoma * with * a * * using * * * * *</de_id>
    </rdf:Description>
    <rdf:Description about="urn:PMID-16160484">
         <dc:title>
unusual morphologic features of endometrial stromal tumor a
         report of 2 cases
         </dc:title>
         <v:autocode term="tumor" code="C0000000" />
<v:autocode term="endometrial stromal tumor" code="C0000000"/>
         <v:autocode term="stromal tumor" code="C6781000" />
<de_id> * * * * of endometrial stromal tumor a * of * * *
   *</de_id>
    </rdf:Description>
    <rdf:Description about="urn:PMID-16160483">
         <dc:title>
pleomorphic adenoma with extensive lipometaplasia report of
         three cases
         </dc:title>
<v:autocode term="pleomorphic adenoma" code="C8602000" />
         <v:autocode term="adenoma" code="C0000000" />
<de_id> * pleomorphic adenoma with * * * of * * * *</de_id>
    </rdf:Description>
    <rdf:Description about="urn:PMID-16160482">
         <dc:title>
         abnormal intrahepatic portal vasculature in native and
         allograft liver biopsies a comparative analysis
         </dc:title>
         <de_id> * * * * * in * and * * * a * * * *</de_id>
    </rdf:Description>
    <rdf:Description about="urn:PMID-16160481">
         <dc:title>
gastrointestinal stromal tumor of the stomach in children
and young adults a clinicopathologic immunohistochemical and
molecular genetic study of 44 cases with long-term follow-
         up and review of the literature
         </dc:title>
         <v:autocode term="tumor" code="C0000000" />
<v:autocode term="tumor of the stomach" code="C3387000" />
<v:autocode term="gastrointestinal stromal tumor of the
   stomach" code="C5806000" />
```

```
        <v:autocode term="stromal tumor" code="C6781000" />
<v:autocode term="gastrointestinal stromal tumor" code="" />
<de_id> * gastrointestinal stromal tumor of the stomach in *
   and * * a * * and * * * of * * with * * and * of the * **</de_id>
    </rdf:Description>
```

8.4. SCRUBBING TEXT USING THE DOUBLET METHOD

Another method for data scrubbing uses a variation of the doublet method. It uses an external list of pre-approved word doublets (about 80,000 of them). The doublet list is chosen to contain no identifying terms. My current list of doublets was derived from two open-source medical vocabularies (MeSH and the Neoplasm Classification, see Appendix). In this simple algorithm, the text to be scrubbed is parsed, and all the doublets in the text that match a term in the approved list are retained. An asterisk replaces everything else. It works fast (1 mbyte per second on my 1.6 GHz CPU) and doesn't allow any unlisted doublets to slip through. It retains more words from the text than the concept match algorithm, thus providing an output that is easier to read.

The value of the use of doublets (instead of approved words) is that many seemingly innocuous words (such as "No") can be a person's name ("Dr. No is in the hospital"). This means that "no" should not be used as a stop word in the Concept-Match method, with every occurrence of the word "no" replaced by an asterisk. In the doublet method, there is no list of stop words. There is only a list of acceptable doublets. Presumably "no tumor" and "no evidence" and "is no" would be included in the list of pre-approved doublets. Occurrences of the word "no" preceding "tumor" and "evidence" and following "is" would be preserved. "No" occurring after the word Doctor or Dr., signifying a person's name, would be omitted from the list of pre-approved doublets and subsequently excluded from the output of the scrubber script. The same benefit would apply to every stop word that can occur in innocuous or dangerous doublets. The doublet scrubber can be scripted in fewer than 20 Perl command lines.

`Scrub.pl` prompts the user to enter some text and produces a "scrubbed" output to the monitor.

The output preserves pre-approved doublets appearing in the submitted text. Everything else is replaced by an asterisk (see List 8.4.2). The same script could easily be modified to accept a text file.

LIST 8.4.1. PERL SCRIPT scrub.pl **DE-IDENTIFIES, SCRUBS, AND PRIVATIZES TEXT**

```
#!/usr/local/bin/perl
open (TEXT, "doubdb.txt")||die"cannot";
   #list of about 80,000 acceptable doublets (see List 8.5.1.)
$line = " ";
while ($line ne "")
   {
   $line = <TEXT>;
   $line =~ s/\n//o;
   $line =~ s/ +$//o;
   $doubhash{$line} = "";
   #creates an associative array of pre-approved doublets
   #from the list contained in the doubdb.txt file
   }
close TEXT;
print "What text would you like scrubbed?\n";
$line = <STDIN>;
$line =~ s/\n//;
$line = lc($line); #convert text to lowercase
$phrase =~ s/\'s//g; #removes possessives
#hodgkin's disease becomes hodgkin disease
$phrase =~ s/\,/ /g;  #removes commas
$line =~ s/[^a-z0-9 \-]/ /g;
   #replaces non-alphanumerics with a space
@hoparray = split(/ +/,$line);
   #creates an ordered array from the text words
for ($i=0;$i<(scalar(@hoparray));$i++)
   #steps through the array of words
   {
$doublet = "$hoparray[$i] $hoparray[$i+1]";
    #finds successive overlapping word doublets
if (exists $doubhash{$doublet})  #checks to see
   #if the doublet is in the set of pre-approved doublets
       {
        print " $hoparray[$i]";
        #prints the first word of the doublet

       $lastword = " $hoparray[$i+1]";
       #saves the second word of the doublet

       }
   else
```

```
    #if the doublet is not in the list of pre-approved doublets
        {
        print $lastword;  #doublet not in database, so print the
        #second word of the last previous matching doublet
        $lastword = " \*";
        #load an asterisk into the variable containing
        #the last matching word
        }
    }
exit;
```

LIST 8.4.2. EXAMPLES OF SCRUBBED TEXT

Basal cell carcinoma, margins involved
Scrubbed text... basal cell carcinoma margins involved

Rhabdoid tumor of kidney
Scrubbed text... rhabdoid tumor of kidney

Mr. Brown has a basal cell carcinoma.
Scrubbed text... * * has a basal cell carcinoma

Mr. Brown was born on Tuesday, March 14, 1985.
Scrubbed text... * * * * * * * * *

The doctor killed the patient.
Scrubbed text... * * * * *

8.5. CREATING A LIST OF SAFE DOUBLETS

The script in section 8.4 used an external file consisting of pre-approved doublets. Pre-approved doublets are two-word terms that do not contain patient names or other identifiers and that consist of words that might appear in any biomedical text, such as "when the, if the, adenocarcinoma of, the tumor." Pre-approved doublets for any scientific domain can be easily and quickly compiled by parsing a domain-specific nomenclature and extracting all the doublets (List 8.5.1).

In the `tiedoub.pl` script, we use the 2007 version of MeSH (Medical Subject Headings, see Appendix) to extract a set of pre-approved dou-

LIST 8.5.1. PERL SCRIPT `tiedoub.pl` **COMPILES DOUBLETS FROM A NOMENCLATURE**

```
#!/usr/local/bin/perl -w
open (TEXT,"d2007.bin")||die"Cannot";
#draws doublets from the 2007 version of MeSH (see Appendix)
$line = " ";
while ($line ne "")
  {
  $line = <TEXT>;
  next unless (($line =~ /^ENTRY \= /)||($line =~ /^MS \= /));
  $line =~ /^[A-Z]+ \= /;
  $phrase = $';
  $phrase = lc($phrase);
  $phrase =~ s/\n//o;
  $phrase =~ s/\'s//g;
  $phrase =~ s/\W/ /g;
  $phrase =~ s/^ +//o;
  $phrase =~ s/ +$//o;
  $phrase =~ s/ +/ /g;
  @hoparray = split(/ /,$phrase);
  for ($i=0;$i<(scalar(@hoparray)-1);$i++)
      {
      $doublet = "$hoparray[$i] $hoparray[$i+1]";
      $doubhash{$doublet}="";
      }
  }
close TEXT;
open (OUT, ">doubdb.txt")||die"cannot";
while ((my $key, my $value) = each(%doubhash))
   {
   if ($key =~ /^[\w]+ [\w]+/)
      {
      print OUT "$key\n";
      }
   }
exit;
```

blets. The MeSH file is `d2007.bin`, the ASCII flat-file version of MeSH. This file exceeds 26 megabytes in length. Each record of the file contains a medical term and several types of annotation. The lines of the MeSH records that contain simple text are preceded by the strings "ENTRY =" or "MS = "

Line 11 of the script is:

```
next unless (($line =~ /^ENTRY \= /)||($line =~ /^MS \= /));
```

This line excludes MeSH lines that do not contain simple text. The remainder of the script converts the text to lowercase, removes extraneous punctuation, and parses out the doublets from the text. The resulting output file, doubdb.txt consists of thousands of doublet terms that contain no patient identifiers.

8.6. REMOVING DUPLICATE ITEMS FROM A LIST FILE

Sometimes, particularly when you have created a file containing a long list, you might inadvertenly add duplicate items to the list. In the case of doubdb.txt, you may wish to add doublet terms that were not included in your original source text. After incrementing a list file over months and years, it is easy to introduce duplicate terms. A simple Perl script can easily remove duplicate items from a list file (see List 8.6.1).

LIST 8.6.1. PERL SCRIPT dup_out.pl REMOVES DUPLICATE LINES FROM A LIST FILE

```
#!/usr/local/bin/perl
open (TEXT, "stopdoub.txt");
$line = " ";
while ($line ne "")
  {
  $line = <TEXT>;
  $getline = $line;
  $getline =~ s/\n$//o;
  next if ($getline != / /);
  $hash{$getline}="";
  }
close TEXT;
use File::Copy;
unlink("stopdoub.txt");
open (HOLD, ">stopdoub.txt");
@keysarray = sort (keys(%hash));
foreach $thing (@keysarray)
   {
   print HOLD "$thing\n";
   }
close HOLD;
exit;
```

The dup_out.pl script will delete duplicate lines from a file. The script introduces Perl's unlink command. The unlink command deletes a file. It requires the built-in File::Copy module.

```
use File::Copy;
unlink("stopdoub.txt");
open (HOLD, ">stopdoub.txt");
```

Once a file is deleted, the same filename can be used as the receptable for the revised list.

8.7. WARNINGS

As data sharing becomes the norm in biology and medicine, the demand for data-sharing algorithms will increase. Two trends in this area of software research are disturbing to the author (List 8.7.1).

LIST 8.7.1. DISTURBING DEVELOPMENTS IN THE DATA-SCRUBBING FIELD

1. The appearance of data-scrubbing applications that run very slowly
2. The proliferation of patent applications in the area of data scrubbing

Some of the data-scrubbing applications in current use are mind-numbingly slow, requiring up to several seconds to scrub a single surgical pathology report (1 to 10 kilobytes).

A slow scrubber will suffice when the task is small and unlikely to require repeated scrubbing passes. More often, datasets are large. Slow software cannot accommodate the terabytes of textual data that are produced daily in the biomedical realm. Once scrubbed, datasets will often require re-scrubbing. Re-scrubbing is necessary because scrubbing transforms and distorts data. Scrubbed data A cannot be merged with scrubbed data B unless data A and data B have been scrubbed by the same algorithm. In practical terms, this means that data sharing projects will require collaborators to re-scrub their original datasets with a common data scrubber. To do all this scrubbing and re-scrubbing, data scrubbers will need to be very fast.

The scrubbers discussed in this chapter scrub text at about 3,000 to 10,000 reports (1 kb) per second on a relatively slow 1.8 GHz CPU. At this rate, all of the surgical pathology reports produced in the U.S. each year (about 25 million) could be scrubbed in about an hour. Given the increasing speed of computers, readers should probably expect scrubbers to perform at this speed or greater.

Also alarming is the proliferation of patent applications covering fundamental processes in de-identification and data scrubbing. A visit to the U.S. Patent and Trademark Web site (http://www.uspto.gov/patft/index.html) indicates that de-identification protocols are an area of great interest for patent applicants.

It would be unfortunate if the well-intentioned act of sharing scrubbed data engendered legal reprisal. History would suggest that clinicians tend to abandon methodologies encumbered by patent royalties (38), (39).

The best way of ensuring an open environment for data scrubbing is by providing opportunities for the publication of data scrubbing methods. Establishing a wide assortment of published data sharing protocols as "Prior Art" will make it difficult for inventors to include fundamental data scrubbing processes in their patent claims.

CHAPTER 9

Finding and Exchanging Data Through the World Wide Web

9.1. BACKGROUND

What does a Web browser actually do? It packages a request, usually in the form of a Web page location, and sends the request via http (hypertext transfer protocol). The Internet, through no help from your browser, routes the request to the server that contains the requested Web page and sends the Web page back to your computer. The browser parses the Web page, looking for HTML formatting instructions, and displays the Web page.

Whether you click on an active link, submit a text query, or fill in parts of a Web form before clicking on a "submit" button, the browser's job is easy: Send the http request, wait for a response, display the returned Web page described by the HTML formatting instructions.

Imagine how simple the process would be if you did not need to display the received Web page. You could bypass the browser entirely. You could send the request via a short Perl script and save the returned HTML message as a file on your hard disk (for later analysis). You also could do a recursive search: extract all of the Web addresses on the returned HTML message; send new requests to each of these addresses; collect the HTML message of each of these; and then repeat the process for each of the Web addresses extracted from each of the returned Web pages. Your hard disk would soon fill with hundreds of thousands of Web pages. At any time, you could activate another Perl script that would search through these pages for terms of interest, creating a distilled document containing only the information relevant to your particular interests.

If you have an insatiable appetite for data, the Web is a wonderful place to play, and Perl allows you to gather large amounts of information very quickly. As a biomedical professional, your best sources for automated searching will likely be PubMed and Crisp.

The purpose of this chapter is to show how Perl can automate Internet queries and collect the resultant data into files.

9.2. RETRIEVING INFORMATION FROM THE INTERNET

Virtually all scripting languages have modules that permit users to retrieve and parse data using Internet protocols. In many cases, these modules are bundled with the standard distributions of the language interpreters. All the user must do is invoke the module at the top of the script, and all the methods contained in the modules become accessible to the script (List 9.2.1).

LIST 9.2.1. EXAMPLE OF A SIMPLE PROGRAM USING LWP (LIBRARY FOR WWW IN PERL)

```
#!/usr/bin/perl
use LWP::Simple;
print (get "http://www.nih.gov");
exit;
```

There are numerous books that explain the intricacies of finding, downloading, parsing, and transforming data residing on the Internet. A good resource is Clinton Wong's *Web Client Programming with Perl* (40).

9.3. RETRIEVING A FILE FROM THE WEB

The script `lwp_get.pl` will pull any file with a specified URL and a size under 630 Megabytes. The example shown is a >400 Mbyte file from the Lawrence Livermore site (List 9.3.1 and 9.3.2).

The meat of the script, `lwp_get.pl`, is contained in four lines of code:

```
use LWP::Simple;
$line = "http\:\/\/image\.llnl\.gov\/image
\/imagene\/4\.6\/data\/human\-4\.6\.xml";
@list = head($line);
```

LIST 9.3.1. PERL SCRIPT `lwp_get.pl` **PULLS FILES FROM URLS**

```
#!/usr/bin/perl
use LWP::Simple;
$line =
    "http\:\/\/image
    \.llnl\.gov\/image\/imagene\/4\.6\/data\/human\-4\.6\.xml";
#provide a file url here
@list = head($line);
print "content type is $list[0]\n";
#@list is a data array provided by the module
print "document length is $list[1]\n";
print "modified time is $list[2]\n";
print "expires is $list[3]\n";
print "server is $list[4]\n";
if ($list[1] < 630000000) #Let's not get greedy
     {
     $file = "bigxml.xml"; #arbitrary filename to accept data
     getstore($line,$file); #invoke the module's getstore method
     }
exit;
```

LIST 9.3.2. OUTPUT OF `lwp_get.pl`

```
c:\ftp>perl lwp_get.pl
content type is text/xml
document length is 427526660
modified time is 1086002464
expires is
server is Apache/1.3.26 (Unix)]
mod_ssl/2.8.9 OpenSSL/0.9.6e mod_perl/1.24 ApacheJser
```

The module is called with the first line. The URL of the desired Web site is provided in the second line. The third line calls the module's "`head()`" method, described in the module's documentation.

The script prints some basic header information for the file, using the LWP::Simple module's `head()` command, which creates an array of information. The `getstore()` command extracts the file and puts it into a provided filename. In this case, the output file is an XML document 427,526,660 bytes in length.

9.4. RSS ACCUMULATORS

RSS is an acronym for Really Simple Syndication (RSS version 2.0), Rich Site Summary (RSS versions 0.91 or 1.0), or RDF Site Summary (RSS versions 0.9 and 1.0). RSS is a rare example of an acronym whose expansion changes with the protocol version. The wildly popular (at the time of this writing) RSS documents, written in a simplified version of RDF, contain links and associated annotations for Web articles.

Content providers (for example, magazines, newspapers, journals) publish RSS feeds on their Web sites. These feeds are XML documents that are typically short-lived and often updated (see Figure 9-1). Users typically acquire software that aggregates information from a list of user-preferred RSS feeds and yields the data as an accessible file. RSS aggregators can be designed to find new articles appearing in any of the feeds.

An example of a simple RSS aggregator searches for text and links from two popular RSS sites: slashdot and PNAS (Proceedings of the National Academy of Sciences) (List 9.4.1).

This Perl script uses two modules: LWP, which comes bundled with Perl; and XML::RSS, which can be easily obtained from CPAN or ActiveState (List 9.4.2).

Figure 9-1 Example of an RSS page.

LIST 9.4.1. PERL SCRIPT `rss.pl`, FOR EXTRACTING INFORMATION FROM RSS FEEDS

```
#!/usr/bin/perl
use XML::RSS;
use LWP::Simple;
$rss = new XML::RSS();
#We'll make a new RSS object that can access
#the RSS module methods
open (STDOUT,">rss_hold.txt");
#We'll put the results in an external file, rss_hold.txt
@rss_array = ("http\:\/\/www\.pnas\.org\/rss\/ahead\.xml",
"http\:\/\/www\.slashdot\.org\/index\.rss");
#We'll use slashdot and PNAS, as example RSS feeds
foreach $value (@rss_array)
  {
  $information = get($value);
  $rss->parse($information);
  #The accessor methods below are described in the RSS module
  foreach $item (@{$rss->{'items'}})
      {
      $title = $item->{'title'};
      $url = $item->{'link'};
      #Mild attempt at prettifying the output
      #You could prepare an output format in html
      print "\*\*\*\n$title\n$url\n";
      }
  }
exit;
```

LIST 9.4.2. A FEW LINES OF OUTPUT FROM `rss.pl`

```
Physics
Surface molecular view of colloidal gelation
http://www.pnas.org/cgi/content/short/0606116103v1?rss=1
***
Commentary
Proline to the rescue
http://www.pnas.org/cgi/content/short/0606106103v1?rss=1
***
```

9.5. CGI SCRIPTS

From the user's point of view, CGI scripts can be perceived as server-side Perl scripts that are called from client Web pages. A CGI script looks just like any other Perl script. The only difference is that it has a `.cgi` suffix and sits in a special Web server directory called `cgi-bin`.

Typically, Web pages call CGI scripts discretely. The Web surfer rarely knows that it's happening. Here is how it is done.

The Web page contains a link (usually in the form of an active link or a "submit" button on a Web form) that sends the name and location of the CGI script, along with the Web server name, as a URL, using the standard http protocol (see Figure 9-2, List 9.5.1 and 9.5.2).

LIST 9.5.1. HTML SOURCE FOR `post.htm` WEB PAGE TO CALL A CGI SCRIPT

```
<html><head><title>post</title>
</head><body>
<br><form name="sender" method="POST"
action="http://THE_URL_OF_THE_SERVER/cgi-bin/regexlho.cgi">
<br><center><input type="text" name="tx" size=38
   maxlength=48 value="">
<input type="submit" name="bx" value="SUBMIT">
<br><br>
</center></form>
<br></body></html>
```

LIST 9.5.2. PERL SCRIPT `regexlho.cgi` SEARCHES A SERVER FILE FOR A WEB CLIENT QUERY

```
#!/usr/bin/perl
print "Content-type: text/html\n\n";
print qq|<html><head><title>Something |;
print qq| more</title></head><body> |;
read(STDIN, $buffer, $ENV{'CONTENT_LENGTH'});
#puts the POST into $buffer
$buffer =~ /tx\=([^&]+)&/;
#assumes that the post take form tx=...&

$regex = $1;
print qq|\n<br>$regex|;
open (TEXT, "dataset.txt")||die"Cannot";
@startval = times();
```

```
$line = " ";
while ($line ne "")
  {
  $line = <TEXT>;
  if ($line =~ /$regex/i)
    {
    $count++;
    if ($count > 30)
      {
      last;
      }
    print qq|\n<br>$line\n<br>|;
    }
  }
close TEXT;
@stopval = times();
$totaltime = ($stopval[0]-$startval[0]);
print qq|\n<br>The total time for script was $totaltime|;
print qq|\n <br><br></body></html> \n\n\n\n |;
exit;
```

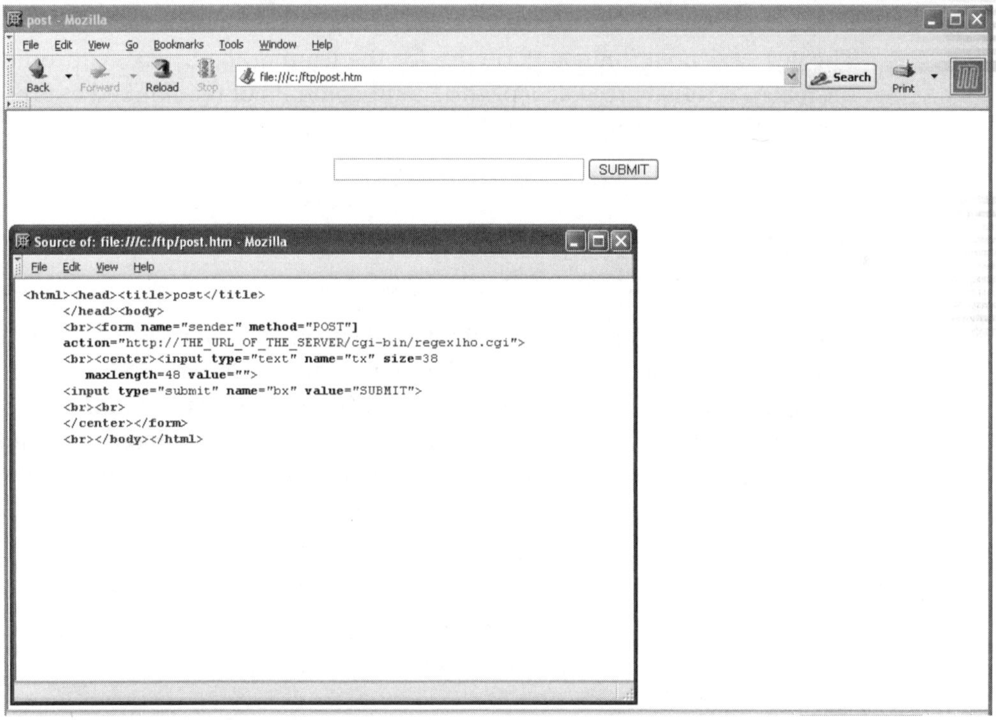

Figure 9-2 A simple Web page that calls a CGI script.

Most CGI scripts are constructed to do a few basic chores (List 9.5.3).

LIST 9.5.3. BASIC TASKS OF A CGI SCRIPT

1. Send print instructions that create a header for an html page.
2. Accept a GET or a POST message from a Web client form.
3. Do something with the GET or POST message. This might involve adding the message contents to a database, or using the message contents as a database query.
4. Send print instructions for the body of text in html format.
5. Send print instructions for the footer of an html page.

In a CGI script, all print commands are sent back to the client. Print commands that constitute a well-formed HTML page will create a response Web page that the client can view. Let us review the regexlho.cgi script. The first three lines send the header of an HTML page to the client's browser:

```
print "Content-type: text/html\n\n";
print qq|<html><head><title>Something |;
print qq| more</title></head><body> |;
```

The qq|...| notation is a Perl cure for the leaning toothpick syndrome. Basically, it tells Perl to accept the literal contents of the flanked expression, without needing to backslash special characters.

```
qq|<html><head><title>Something |
```

is the same as:

```
"\<html\>\<head\>\<title\>Something "
```

The regexlho.cgi script accepts a POST message from a Web client's post.htm Web page (List 9.5.1). The next line reads the POST message into the variable *$buffer. This will be explained in more detail in section 9.6.* The $buffer string is trimmed of extraneous characters passed by the post.htm page, including the characters "tx=", containing the name of the input specified in post.htm.

```
read(STDIN, $buffer,$ENV{'CONTENT_LENGTH'});
#puts the POST into $buffer
$buffer =~ /tx\=([^&]+)&/;
#assumes that the post take form of tx= ... &
$regex = $1;
#puts test in a variable to be used in a pattern match
```

The next lines describe a simple file search for matches to the submitted text.

Notice that the print command (sending data to the client) is in the form of a formatted HTML line, with flanking line breaks enclosing a variable Perl will automatically interpret.

```
print qq|\n<br>$line\n<br>|;
```

The final lines close the HTML page by closing the HTML body tag and the document's HTML tag.

```
print qq|\n <br><br></body></html> \n\n\n\n |;
exit;
```

The returned output is an HTML page viewed on the client's browser. The page would have an appearance much like Figure 16-2 (Chapter 16, Section 7).

9.6. SENDING AND RECEIVING `POST` AND `GET` COMMANDS

Web clients send data to server-side CGI scripts through two mechanisms: `GET` and `POST`.

Typically, Web forms specify a `GET` or `POST` method as an attribute to the form element.

For example:

```
<br><form name="sender" method="GET">
```

Or:

```
<br><form name="sender" method="POST">
```

If the form designates that the entered data is to be sent as a `GET` message, the CGI script receives a fragment of text in the *$ENV{'QUERY_STRING'}* environment variable.

A typical line in a CGI script that accepts a client's Web form query term is:

```
$query_term = ($ENV{'QUERY_STRING'});
```

The incoming `GET` message is sometimes obfuscated by a variety of special characters inserted by the browser.

I typically massage the `GET` message through several `regex` filters to normalize the text into simple ASCII:

```
$received_string = ($ENV{'QUERY_STRING'});
$received_string =~ s/\+/ /;
$received_string =~ s/\%2B/+/;
```

```
    $received_string =~ s/%(..)/pack("c",hex($1))/ge;
    $received_string =~ /tx
    \=([\d\"\'\(\)\-        \|\[\]\{\}\@\#\!\^\*\_\?\>\<\~\`\"\:\;\\\/\.
    \w\+]+)[\W]*/;
  $received_string = $1;
```

Readers may use their own methods for filtering GET messages based on the particular requirements of their client-server exchanges. If the form designates that the entered data is to be sent as a POST message, the CGI script receives a fragment of text in the <STDIN> file handle.

A typical line that accepts a POST text message from a client Web form is:

```
read(STDIN, $buffer, $ENV{'CONTENT_LENGTH'});
```

Here, the CGI script uses the read() function. The length of the POST is received in the Perl environmental variable *$ENV{'CONTENT_LENGTH'}*. The read statement extracts a string from the STDIN file handle of length equal to the length of the POST message and puts the string into the *$buffer* variable. Aside from the different ways that a CGI script accepts GET and POST data, there are also client-side differences between GET and POST. When a client form sends input text through a GET message, the content of the GET message is appended to the CGI's URL and displayed in the browser location box. As a general rule, GET messages are short (under 1 kbyte), and POST messages are long.

9.7. SECURITY CONSIDERATIONS

CGI programming is not particularly safe for either the client or the server. Clients can include intrusive code within GET or POST messages. This is usually accomplished through an eval statement inserted as the content of the GET or POST, which tells Perl to execute a character string. When the CGI script receives the client data, Perl might innocently evaluate the client's input as though it were a command line. Obviously, a CGI program that does not guard against this and other deceptions might be vulnerable to client attack. Servers can also hurt clients. A CGI script can be programmed to send countless annoying HTML pages to clients or to use the server-client connection for other devious purposes. Anyone using CGI should be aware of the risks.

9.8. CGI PROGRAMMING AS A MODEST INTRODUCTION TO DISTRIBUTED COMPUTING

The only drawback in CGI programming is that you must have access to a server's cgi-bin. This book emphasizes self-reliance and open-source solutions. We have tried to avoid solutions that require proprietary software, proprietary hardware, or complex protocols. In Section 16.6, we will show that every networked computer (including your own) can be a server.

LIST 9.8.1. THE ADVANTAGES OF CGI PROGRAMMING

1. The computations are performed on the server side, not the client side. A Web surfer can blithely send a request (via a CGI calling Web form) that initiates a massive server-side computational effort. A fraction of a second later, a calculated reply appears on her screen.
2. CGI accepts input from the client. This is done through `GET` or `POST` messages.
3. The data resources are stored on the server side, not the client side. A Web request might require the participation of an enormous database or a federation of databases distributed throughout the world. A server-side CGI script can implement and coordinate complex database queries.
4. A server-side CGI script can be easily modified and improved. There is no need to change the client-side Web pages that call the CGI script.

CHAPTER 10

Creating, Parsing, and Transforming XML and Tagged Data

10.1. BACKGROUND

XML is an informatics technology that allows any data element (for example, a gene sequence, the weight of a patient, a biopsy diagnosis) to be bound to the data that describe the data element (that is, the metadata). It is surprising that this simple relationship between data and the data that describes data is the most powerful innovation in information organization since the invention of the book. The way that we choose to implement metadata annotation might have evolved through any number of different architectures. However, XML has emerged as the dominant technology for this purpose (41), (42). Seldom does a new technology arise with all the techniques required for its success, but this seems to be the case for XML.

The purpose of this chapter is to explain why XML is important in biomedicine and to show how Perl scripts can easily parse and extract data from XML documents.

10.2. XML BASICS

In XML, data descriptors (known as XML tags) enclose the data they describe with angle brackets:

```
<birthdate>September 19, 1950</birthdate>
```

`<birthdate>` is the XML tag. The tag and its end tag enclose a data element, which in this case is the unabbreviated month, beginning with an uppercase letter, followed by lowercase letters, followed by a space, followed by a two-digit numeric for the date of the month, followed by a comma and space, and then followed by the four-digit year. The XML tag could have been defined in a separate document detailing the data format of the data element described by the XML tag. ISO-11179 (see Glossary) is a standard that tells people how they should specify the properties of metadata (43). In this case, the metadata is the XML tag <birthdate>. If we had chosen to do so, we could have broken the <birthdate> tag into its constituent parts:

```
<birthdate>
<month_of_birth>September</month_of_birth>
<month_day_of_birth>19<month_day_of_birth>
<year_of_birth>1950<year_of_birth>
</birthdate>
```

XML is powerful because it explains the meaning of data and because it permits us to reach data anywhere on the Internet.

XML achieves its power through six extraordinary properties (List 10.2.1).

An XML file is well formed if it conforms to the basic rules for XML file construction recommended by the W3C (World Wide Web Consortium). This means that it must be a plain-text file, must contain a header indicating that it is an XML file, and must enclose data elements with metadata tags that declare the start and end of the data element. The tags must

Figure 10-1 XML is the father of all metadata.

LIST 10.2.1. SIX EXTRAORDINARY PROPERTIES OF XML

1. Enforced and defined structure (XML rules and schema)
2. Formal metadata (through ISO11179 specification)
3. Namespaces (permits sharing of uniquely identifiable common data elements [CDEs])
4. Linking data via the Internet (through Unique Resource Identifiers [URIs])
5. Logic and meaning (the Semantic Web and Ontologies)
6. Self-awareness (software agents [see Glossary], artificial intelligence [see Glossary], embedded protocols and commands)

conform to certain rules (for example, alphanumeric strings without intervening spaces) and must also obey the rules for nesting data elements.

Most browsers will parse XML files, rejecting files that are not well formed. Having the ability to ensure that every XML file conforms to basic rules of metadata tagging and nesting makes it possible to extract XML files as sensible data structures.

A well-formed XML file must be structured according to either a DTD (Document Type Definition) or to a schema before it can be considered a valid XML document. DTDs and schemas are blocks of descriptors that specify the structure and content of an XML file. Schemas are more popular than DTDs. A variety of schema languages are available (44). Regarding schemas, when an XML file is described by a schema and parsed by a validating browser (or by a so-called XML parser), you can be certain that the data contained in the file conforms to a specified structure and content. Files using the same schema will have the same data organization, and this greatly facilitates data integration between files.

Examples of widely adopted schemas from the field of bioinformatics are MIAME (minimal information for the annotation of a microarray experiment) (45) and MIAME-compliant MAGE-ML (microarray gene expression mark-up language) (46).

The concept of formalized metadata is simple. Consider the seemingly obvious metadata tag `<date>`. Does this designate a day in a calendar, does it represent a type of fruit, or does it refer to a social event?

The International Standards Organization has created a standard way to define metadata. This standard, the ISO 11179, specifies that metadata should have a qualified name or identifier, an authority that registers the name, a versioning history (allowing for modifications), a language or origin, a statement relating to usage, a data-typing statement, and a definition that is unambiguous (43). XML files should always include pointers (that is, links) to the Web addresses of the files that contain the definitions of the metadata tags appearing in the XML file.

Standard methods of organizing and describing data are essential. As an example, the creators of the Tissue MicroArray (TMA, see Glossary) Data Exchange Specification provide a file that lists each of the 80 XML tags used in the specification, along with the ISO 11179 descriptors (List 10.2.2) for each tag (47). This metadata definition file for the TMA data exchange specification (see Glossary) resides at:

http://www.pathology.pitt.edu/pdf/cpctr/tma_cde.htm

LIST 10.2.2. DESCRIPTORS FOR THE METADATA TAG `<core_organism>`

```
core_organism
Identifier: core_organism
Version: 1.0
Registration Authority: Association for Pathology Informatics
Language: English (en)
Obligation: Optional
Datatype: Character String representing taxonomy.dat
   identifier number followed by an allowable taxonomy.dat name
   for the identifier number
Maximum Occurrence: Unlimited
Definition: Organism name at species level for organism
   whose tissue is represented in the donor block
Comment: URL for taxonomy.dat is
   ftp://ftp.ebi.ac.uk/pub/databases/taxonomy/taxonomy.dat The
   correct entry for human tissue is "9606 human"
```

Without fully defined metadata, XML data has no meaning or value.

Problems arise when data from different documents are merged. How can we be sure that XML tags in one document will always mean the same thing when found in another document? This is achieved with namespaces. Whenever you use an element taken from another XML document, you should declare the namespace origin of the element, for example:

```
<table xmlns="http://www.w3.org/TR/html4/">
```

The `xmlns` attribute indicates that the `<table>` element is the same element that is described for the World Wide Web organization's HTML recommendation, and provides the namespace specific to this meaning of the

`<table>` element. Thus `<table>`, as it is used in the data element, has a meaning that cannot be confused with "kitchen table" or "periodic table."

If we had multiple elements from the HTML specification, we might have chosen to list the namespaces near the top of the XML document and assigned a prefix specific to elements belonging to the HTML namespace. Consider this snippet of XML from the Gene Ontology specification (48). Two different namespaces (:go and :rdf) are declared using the xmlns (XML namespace) attribute:

```
<go:go xmlns:go="http://www.geneontology.org/dtds/go.dtd#"
xmlns:rdf="http://www.w3.org/1999/02/22-rdf-syntax-ns#">
<go:version timestamp="Wed Jun 19 12:34:48 1974" />
<rdf:RDF>
<go:term
   rdf:about=
   "http://www.geneontology.org/go#GO:0003673"n_associ
   ations="0">
   <go:accession>GO:0003673</go:accession>
<go:name>Gene_Ontology</go:name>
</go:term>
```

The namespace designations appear later in the XML file as prefixes to the tags defined in the namespace documents (go:term, rdf:about, go:name, go:version, rdf:RDF)

In like manner, a single XML file can use a single metadata tag to mean many different things, so long as each use of a metadata element is prefixed with the correct namespace. The namespace prefix annotation is a powerful informatics tool, permitting XML creators to recycle metadata from different namespaces. When many different researchers use the same metadata, the contained data can often be shared between their databases.

We will revisit ISO-11179 in Chapter 10 Section 8, where we discuss the Dublin Core Common Data Elements.

10.3. COLLECTING THE METADATA ELEMENTS FROM MULTIPLE XML FILES

Sometimes the best way to understand an XML file is to collect the tags (metadata terms) found in the file. Once you have a chance to see the metadata included in a file, it might be useful to extract the data associated with some of the tags of greatest interest to you. If you're trying to combine data from multiple XML files conforming to the same XML schema, it might be convenient to have a script that parses all the files, collecting metadata/data pairs along the way.

The script get_tag.pl parses all the XML files included in the open-access collection of journal articles made available from BioMedCentral. All the files are XML and all the files are structured the same way, using the same metadata tags, making data collection very easy. In addition, the script pulls e-mail addresses and text from the articles, formatting the text as it goes along (List 10.3.1). The script can be adapted for any collection of XML files.

LIST 10.3.1. PERL SCRIPT get_tag.pl COLLECTS XML TAGS FROM AN XML FILE

```
#!/usr/bin/perl
use XML::Parser;
my $parser = XML::Parser->new( Handlers => {
  Init => \&handle_doc_start,
  Final => \&handle_doc_end,
  Start => \&handle_elem_start,
  End => \&handle_elem_end,
  Char => \&handle_char_data,
  });
my $line = " ";
my%filetags;
my $state = 0;
my $count = 0;
my $text = '';
@filelist =  glob('*.xml');  #see glob in Glossary
open(OUT,">tags.out");
foreach $filename (@filelist)
{
$count++;
print "$filename\n";
print OUT "$filename\n";
#last if ($count > 100);
$parser -> parsefile($filename);


sub handle_doc_start
  {
  }


sub handle_doc_end
  {
  }
```

```
sub handle_elem_start
  {
  my ($expat, $name,%atts) = @_;
  #the module passes an array @_ to the subroutine
  $filetags{$name}++;
  if ($name eq "e-mail")
    {
    $state = 1;
    }
  if ($name eq "bdy")
    {
    $state = 2;
    }
  #print "$name\n";
  }


sub handle_elem_end
  {
  my($expat, $name) = @_;
  if ($name eq "e-mail")
   {
   $state = 0;
   $text = "";
   }
  if ($name eq "bdy")
   {
   $state = 0;
   $text =~ s/\n/ /g;
   $text =~ s/[^\d\w\.\,\"\'\$\%\(\)\[\]\{\}\<\>]/ /g;
   $text =~ s/ +/ /g;
   $text =~ s/^ +//g;
   $text =~ s/ +$//g;
   while ($text =~ /.{1,70}[\-\,\. ]/g)
     {
     print OUT "$&\n";
     }
   print "\n";
   $text = "";
   }
  }
```

```
sub handle_char_data
  {
  my $expat = $_[0];
  if ($state == 1)
    {
    $text = "$_[1]";
while ($text =~
   /\b([0-9a-zA-Z\_\.]+\@[0-9a-zA-Z\_\.]+)\b/gm)
      {
      print "$1\n";
      }
    }
  if ($state == 2)
    {
    $text .= "$_[1]";
    }
  }



}


while (($key, $value) = each(%file tags))
    {
    print "$key => $value\n";
    }
print "The total number of files parsed is $count\n";
exit;
```

`Get_tag.pl` calls the `XML::Parser` module that is bundled in the Perl distribution:

```
use XML::Parser;
```

`XML::Parser` is one of the earliest XML parsers written for Perl. There are two types of XML parsers: DOM parsers and event parsers. DOM parsers build, in memory, a data structure equivalent to the document's object model (that is, its DOM). To build a DOM structure, the parser must traverse the entire file, building the relationships between the parent and child nodes of the XML file. Once the DOM data structure is built, the parser can traverse the data structure to find metadata tags and can locate associated data fields that are referenced within the DOM data structure. DOM parsers are slow, and cannot provide any output until the entire XML file has been traversed and all the metadata nodes have been assigned a place in the growing DOM data structure. Most important, DOM parsers simply cannot handle large XML files (hundreds of megabytes or gigabytes). Unless your computer is loaded with lots of memory, DOM parsers will not work for your large, data-mining projects.

Event parsers are fast. As they read a file, events are rapidly triggered:

- Document starts.
- A metadata begin tag is encountered.
- Data is encountered.
- A metadata end tag is encountered.
- Document ends.

`XML::Parser` uses the following syntax to create the event subroutines:

```
my $parser = XML::Parser->new( Handlers => {
Init => \&handle_doc_start,
Final => \&handle_doc_end,
Start => \&handle_elem_start,
End => \&handle_elem_end,
Char => \&handle_char_data,
});
```

When the parser encounters the beginning of a tag, it automatically passes a list of parameters to the subroutine in the @_ array variable.

To understand the `get_tag.pl` script, it is not necessary to know all of the passed items. It is sufficient to note that the second item (array item 1 when you start counting at zero) is the name of the XML tag. Once you have the name, you can program all manner of actions. You can use the metadata name as a key in a hash and increment the value of the hash item by one at each event in which the name of the tag is encountered (thus counting the occurrence of the tag in the XML mile). You can create conditional statements that change the action of other script subroutines based on which metatdata tag is encountered in the current event:

```
sub handle_elem_start
  {
  my ($expat, $name, %atts) = @_;
  $filetags{$name}++;
  if ($name eq "e-mail")
    {
    $state = 1;
    }
  if ($name eq "bdy")
    {
```

```
      $state = 2;
    }
  #print "$name\n";
  }
```

XML::Parser is a simple, convenient module for parsing large or multiple XML files and initiating processes determined by which metadata tags are encountered.

10.4. CONVERTING POD TEXT TO AN INDEXED HTML FILE

A common task for informaticians is transforming text to a tagged format, such as HTML or XML. This is most easily done when the text is highly structured. Perl contains its own language for encapsulating text within a Perl script. It is called POD (plain old documentation). The POD text within a Perl script or module contains information that describes the script and explains how the script should be implemented. It is extremely easy to convert embedded POD text to HTML (List 10.4.1 and 10.4.2). The resulting HTML document can be viewed on a browser.

LIST 10.4.1. SAMPLE OF POD SYNTAX FROM THE CLASS MODULE

Parse.pm

```
=head1 NAME
Parse - Parent class for Encode, Sentence, and Relation


=head1 VERSION


Version 1.00 released October 4, 2001


=head1 USAGE


The class packages (Parse.pm, Sentence.pm, Encode.pm,
   Relation.pm)
should all be placed together in a subdirectory. They do not
   need to
be placed in a Perl Lib subdirectory.
```

```
    You can let Perl know where the class packages are by
       including the
    path to the packages in a call to the Lib module, as shown in
the following
    for a Windows subdirectory system. In this example, the class
    packages would all be inside the parse subdirectory.


    use Lib ("c:\\newinfo\\parse\\");


    The use Lib command should be near the top of any script you
       write, together
    with any of the other pragma commands you might use.


    =head1 DESCRIPTION


    The Parse class packages consist of the parent class
       Parse.pm and the
    three subclasses Encode.pm, Sentence.pm, and Relation.pm.


    =head2 OVERVIEW


    The main purpose of the set of Perl packages is to annotate
       medical
    text with UMLS codes or with expansions of abbreviations or
       to find
    relationships between concepts found within the text.


    The main purpose of Parse.pm is to provide a new constructor
       for all
    the subclasses (Encode.pm, Relation.pm, and Sentence.pm) that
       ensures
    that all the subclasses have access to the same superclass hash.
```

LIST 10.4.2. PERL SCRIPT `pod2htm.pl` CONVERTS PLAIN OLD DOCUMENTATION (POD) TO HTML

```
#!/usr/local/bin/perl
open(STDOUT,">pod_out.htm");
use Pod::Html;
$filename = "xmlparse.pl";
pod2html($filename);
close STDOUT;
exit;
```

Figure 10-2 shows the HTML output for the POD document in List 10.4.1.

10.5. CONVERTING AN EXCEL FILE TO AN XML FILE USING `XML::Excel`

This short (five-line), but powerful Perl script uses the `XML::Excel` module that can be downloaded from the `ActiveState ppm` service. All you do is declare an Excel object from any `.xls` file (in this case, `trial2.xls`). The headings => 1 attribute tells the script to use the first line of the Excel spreadsheet (the column names) as the XML tags in the output file.

LIST 10.5.1. PERL SCRIPT `xl2xml.pl` CONVERTS AN EXCEL FILE TO XML

```
#!/usr/bin/perl
use XML::Excel;
$file = "trial2.xls"; #an Excel file
$excel_obj = XML::Excel->new();
$status = $excel_obj->parse_doc($file, {headings => 1});
$excel_obj->print_xml("trial2.xml");
exit;
```

At the Wellcome Trust Sanger Institute site, you can download an Excel file that is the latest version of the census of human cancer genes (49), (50).

http://www.sanger.ac.uk/genetics/CGP/Census/

The Excel file of the human cancer gene census can be downloaded by clicking "Complete working `list.xls`". The downloaded file should be re-named `onco.xls`. A Perl script converts the Excel file, `onco.xls`, into an XML object (see List 10.5.2). There is a problem, however, and the Excel file cannot be directly converted to proper XML. This is because some of the headers of the Excel columns contain spaces and do not conform to valid XML tags.

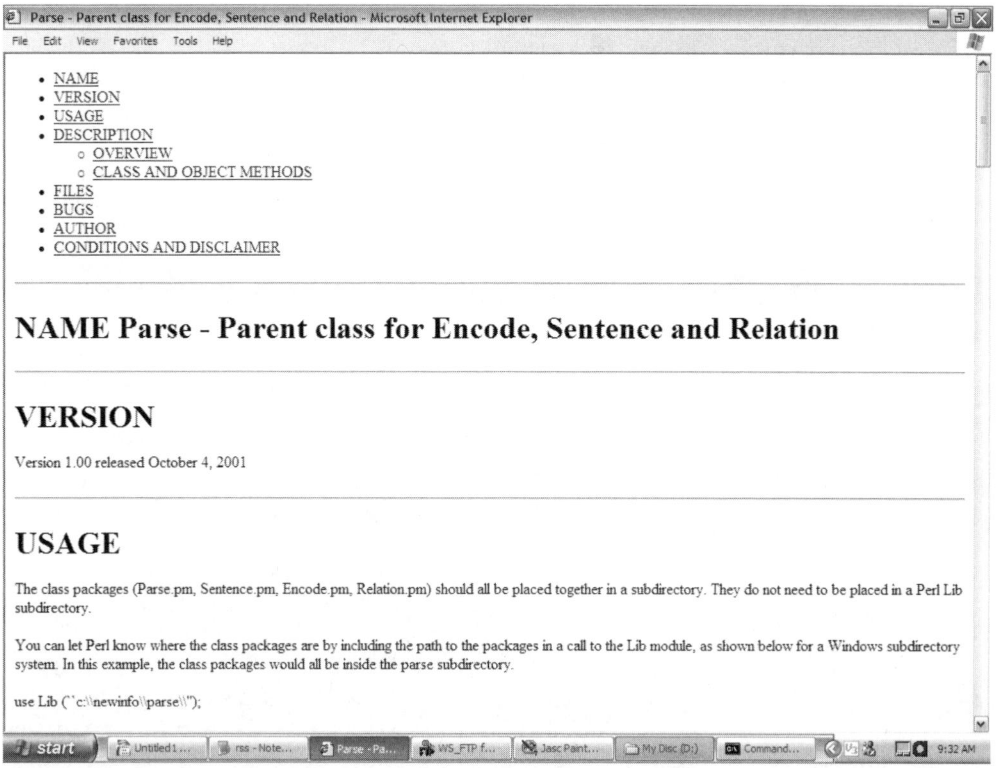

Figure 10-2 A typical browser screen of a POD file transformed to HTML using the Pod::Html module.

The output file produced by the XML::Excel module needs to be parsed again, transforming non-conformant tags to proper XML. A sample output record is shown:

```
<record>
<symbol>ABL1</symbol>
<name>v-abl Abelson murine leukemia viral oncogene homolog
1</name>
<locuslink_id>25</locuslink_id>
<protein_id_swissprotrefseq>P00519</protein_id_swissprotref-
seq>
<chr>9</chr>
<chr_band>9q34.1 </chr_band>
<cancer_somatic_mut>yes</cancer_somatic_mut>
<cancer_germline_mut></cancer_germline_mut>
<tumor_types_somatic_mutations>CML,ALL</tumor_types_somatic_m
utations>
<tumor_types_germline_mutations></tumor_types_germline_muta-
tions> <cancer_syndrome></cancer_syndrome>
<tissue_type>L</tissue_type>
```

```
<cancer_molecular_genetics>Dom</cancer_molecular_genetics>
<mutation_type>T</mutation_type>
<translocation_partner>BCR, ETV6, </translocation_partner>
<other_germline_mut></other_germline_mut>
<other_syndrome_or_disease></other_syndrome_or_disease>
</record>
```

LIST 10.5.2. PERL SCRIPT `xmlcensu.pl`**, CONVERTS THE GENE CENSUS EXCEL FILE TO XML**

```
#!/usr/bin/perl
$file = "onco.xls";
use XML::Excel;
$excel_obj = XML::Excel->new();
$status = $excel_obj->parse_doc($file, {headings => 1});
$excel_obj->print_xml("oncogene.one");
open (TEXT,"oncogene.one");
open (OUT,">oncogene.xml");
$line = " ";
while ($line ne "")
   {
   $line = <TEXT>;
   $line =~ s/\/refseq/refseq/ig;
   $line =~ s/syndrome\/disease/syndrome or disease/ig;
   while ($line =~ /\<([^\>]+)\>/g)
     {
     $front = $`;
     $back = $';
     $middle = $1;
     $middle =~ s/^ +//o;
     $middle =~ s/ +$//o;
     $middle =~ s/ +/\_/g;
     $middle = lc($middle);
     $middle =~ s/[^a-z0-9\_\/]//g;
     #$line = $front . "\<" . $middle . "\>" . $back;
     $line = $front . "\*" .  $middle . "\%" . $back;
     }
   $line =~ s/\*/\</g;
   $line =~ s/\%/\>/g;
   $line =~ s/\ /   /g;
   $line =~ s/[\<]+/\</g;
   print OUT $line;
   }
exit;
```

CHAPTER 11

Metadata, Ontologies, and the Meaning of Everything

11.1. BACKGROUND

Many of us in the health-care and life sciences spend a significant portion of our professional lives preparing data that will be entered into large databases. We are assured that the entered data will be available to us and will support an infinite variety of research projects. At first, we are happy. The databases can retrieve cases of interest and can group records by any of the data elements contained in the records. Summary data is easy to generate from databases. Readily available software can create visually impressive charts and graphs from the data summaries.

Databases sometimes disappoint us. For a variety of reasons, we can never seem to merge our data with data produced by colleagues who use a different database. Often, databases within a single institution are incompatible. The solution, we are told, is data standards. If we could mutually decide what data we need to store and how the stored data should be structured, then our databases would be compatible. Our trust in the wisdom of standards has fueled thousands of standards initiatives in the healthcare field.

The problem is that compliance with standards is often very low, and standards themselves can be flawed. As technologies change, standards do not always keep pace. This often results in obsolete standards or standards with multiple versions that in turn have idiosyncratic implementations.

RDF (Resource Description Framework) is a formal method for describing specified data objects with paired metadata and data. We will see that RDF-specified data provides some of the functionality of standards. In addition, RDF specifications greatly expand our ability to understand information. In my opinion, all life-science professionals can benefit from understanding the basics of RDF.

The purpose of this chapter is to explain the importance of RDF and to show how Perl scripts can facilitate the integration of heterogeneous biomedical data using RDF triples. In Chapter 18, we will describe a project where RDF and Perl are used to implement a medical image data specification.

11.2. THE FAILURE OF XML SCHEMA

XML represents an enormous advance over traditional methods for organizing data. Attaching metadata to data values provides an easy way of representing information. XML data can be structured with XML schemas that list the metadata elements that must be included in conforming documents, along with their relative locations in the document. The inclusion of data typing within the XML schema language provides a way of ensuring that the data contained in an XML document will satisfy specified requirements in the XML schema.

The problem with XML schema is that it functions within the realm of data structure, not data meaning. XML schema tells you how to organize your data, but it does not tell you the object to which your data refers. Data meaning is achieved when a metadata-data pair refers to a specific object. More simply put, data is only meaningful when it describes something. XML schemas only provide the structure for the occurrences of metadata-data pairs, and this is insufficient to understand a document. If you need documents that contain meaningful assertions, you must move beyond XML and into the world of RDF.

11.3. THE VALUE OF RDF

RDF (Resource Description Framework) is a semantic extension of XML that achieves meaning from data by representing all information as data triples (51). RDF triples consist of a unique data object and a metadata-data pair that belongs to the unique data object. RDF triples for unique objects can be collected from many different XML databases and can be merged without loss of meaning. RDF triples provide an opportunity for conducting advanced data analyses on grouped unique objects that share common properties (via ontologies).

11.4. STATEMENTS OF MEANING

In informatics, assertions have meaning whenever a pair of metadata and data (the descriptor for the data and the data itself) is assigned to a specific subject.

Figure 11-1 The meaning of everything: a unique object containing a metadata descriptor enclosing a data container.

Triples consist of: Specified subject then metadata then data

Some triples found in a medical dataset:

"Jules Berman" "blood glucose level" "85"

"Mary Smith" "blood glucose level" "90"

"Samuel Rice" "blood glucose level" "200"

"Jules Berman" "eye color" "brown"

"Mary Smith" "eye color" "blue"

"Samuel Rice" "eye color" "green"

Some triples found in a haberdasher's dataset

"Juan Valdez" "hat size" "8"

"Jules Berman" "hat size" "9"

"Homer Simpson" "hat size" "9"

"Homer Simpson" "hat type" "bowler"

Triples collected from both datasets whose subject is "Jules Berman"

"Jules Berman" "blood glucose level" "85"

"Jules Berman" "eye color" "brown"

"Jules Berman" "hat size" "9"

Triples can port their meaning between different databases because they bind described data to a specified subject. This supports data integration of heterogeneous data and facilitates the design of software agents. A software agent, as used here, is a program that can interrogate multiple RDF documents on the Web, initiating its own actions based on inferences yielded from retrieved triples. RDF (Resource Description Framework) is a syntax for writing computer-parsable triples. For RDF to serve as a general method for describing data objects, we need to answer four questions (List 11.4.1).

LIST 11.4.1. FOUR FUNDAMENTAL QUESTIONS RELATED TO RDF DATA DESCRIPTIONS

1. How does the triple convey the unique identity of its subject? In the triple, "Jules Berman", "blood glucose level" "85", the name "Jules Berman" is not unique and might apply to several different people.
2. How do we convey the meaning of metadata terms? Perhaps one person's definition of a metadata term is different from another person's. For example, is "hat size" the diameter of the hat, the distance from ear to ear on the person who is intended to wear the hat, or a digit selected from a predefined scale?
3. How can we constrain the values described by metadata to a specific data type? Can a person have an eye color of 8? Can a person have a blood glucose value of chartreuse?
4. How can we indicate that a unique object is a member of a class and can be described by metadata shared by all the members of a class?

Much of the remainder of this chapter will be devoted to answering these four questions.

11.5. RDF TRIPLES

RDF is a specialized XML syntax for creating computer-parsable files consisting of triples. The subject of the RDF triple is invoked with the `rdf:about` attribute. Following the subject is a metadata-data pair.

Let us create an RDF triple whose subject is the jpeg image file specified as: *http://www.gwmoore.org/ldip/ldip2103.jpg.* The metadata is `<dc:title>` and the data value is "Normal Lung".

```
<rdf:Description
rdf:about="http://www.gwmoore.org/ldip/ldip2103.jpg">
<dc:title>Normal Lung</dc:title>
</rdf:Description>
```

Done! We have created our first RDF triple. Now, let us look at three triples in proper RDF syntax:

```
<rdf:Description
rdf:about="http://www.gwmoore.org/ldip/ldip2103.jpg">
<dc:title>Normal Lung</dc:title>
</rdf:Description>
<rdf:Description
rdf:about="http://www.gwmoore.org/ldip/ldip2103.jpg">
<dc:creator>Bill Moore</dc:creator>
</rdf:Description>
<rdf:Description
rdf:about="http://www.gwmoore.org/ldip/ldip2103.jpg">
<dc:date>2006-06-28</dc:date>
</rdf:Description>
```

Each of the triples is about the same unique object (a URL for a jpeg image). RDF permits us to collapse multiple triples that apply to a single subject. The following RDF:Description statement is equivalent to the three prior triples:

```
<rdf:Description
rdf:about="http://www.gwmoore.org/ldip/ldip2103.jpg">
<dc:title>Normal Lung</dc:title>
<dc:creator>Bill Moore</dc:creator>
<dc:date>2006-06-28</dc:date>
</rdf:Description>
```

An example of a short but well-formed RDF image specification document containing our triples is the following:

```
<?xml version="1.0"?>
<rdf:RDF
xmlns:rdf="http://www.w3.org/1999/02/22-rdf-syntax-ns#"
xmlns:dc="http://purl.org/dc/elements/1.1/">
<rdf:Description
rdf:about="http://www.gwmoore.org/ldip/ldip2103.jpg">
<dc:title>Normal Lung</dc:title>
<dc:creator>Bill Moore</dc:creator>
<dc:date>2006-06-28</dc:date>
</rdf:Description>
</rdf:RDF>
```

The first line tells you that the document is XML. The second line tells you that the XML document is an RDF resource. The third and fourth lines invoke the namespace documents that are referenced within the document (more about this later). Following that comes the first RDF statement. Believe it or not, we have just covered 95 percent of what you need to know to specify data with RDF.

11.6. EXTRACTING TRIPLES FROM AN RDF DOCUMENT

As you must have guessed, it is easy to extract triples from any RDF document (List 11.6.1).

The `rdf2trip.pl` script is short but incredibly useful. It creates an output that can be used in an expanded script to collect and merge triples from multiple documents. The script uses two modules: `Data::Dumper` and `RDF::Core`. `Data::Dumper` is included in the standard Perl distribution. Data dumper, as its name suggests, dumps out the contents of a variable that contains (or points to) any Perl data structure. We will not cover References (also known as Pointers) in this book. Suffice it to say that many Perl modules return complex data that is referenced by special variables. `Data::Dumper` takes such variables and displays the contained data. We will use `Data::Dumper` again in Sections 13.11 and 15.3.

`RDF::Core` is an easily accessible module, available from the ActiveState package manager. We use the Parser module within the `RDF::Core` module. The syntax for calling the module is `RDF::Core::Parser`.

The module delivers triple variables with the following names and meanings:

```
subject_ns subject_name subject_uri predicate_ns
predicate_name predicate_uri object_ns object_name object_uri
```

LIST 11.6.1. PERL SCRIPT `rdf2trip.pl` EXTRACTS TRIPLES FROM RDF

```
#/usr/local/bin/perl
use Data::Dumper;
use RDF::Core::Parser;


my %options = (Assert => \&handleAssert,
               BaseURI => ".",
               BNodePrefix => "id");
my $parser = new RDF::Core::Parser(%options);
$parser->parseFile('image.rdf');
#print Dumper($parser);
sub handleAssert
  {
  print Dumper(@_);
  print "\n\n";
  }


exit;
```

In this book, the term "predicate" is used interchangeably with the metadata of the triple, or the key to the key-value pair describing the subject. The object of the triple is the data value of the triple or the value of the key-value pair that describes the subject.

LIST 11.6.2. EXCERPTED OUTPUT OF `rdf2trip.pl`

```
C:\ftp>perl rdf2trip.pl


$VAR1 = 'predicate_uri';
$VAR2 = 'http://www.someplace.org/rdf-image#camera';
$VAR3 = 'subject_uri';
$VAR4 = 'http://www.gwmoore.org/ldip/ldip2201.jpg';
$VAR5 = 'predicate_ns';
$VAR6 = 'http://www.someplace.org/rdf-image#';
$VAR7 = 'object_lang';
$VAR8 = '';
$VAR9 = 'predicate_name';
```

```
$VAR10 = 'camera';
$VAR11 = 'object_datatype';
$VAR12 = '';
$VAR13 = 'object_literal';
$VAR14 = 'yes';



$VAR1 = 'predicate_uri';
$VAR2 = 'http://www.someplace_other.com/rdf-otherimage#camera';
$VAR3 = 'subject_uri';
$VAR4 = 'http://www.gwmoore.org/ldip/ldip2201.jpg';
$VAR5 = 'predicate_ns';
$VAR6 = 'http://www.someplace_other.com/rdf-otherimage#';
$VAR7 = 'object_lang';
$VAR8 = '';
$VAR9 = 'predicate_name';
$VAR10 = 'camera';
$VAR11 = 'object_datatype';
$VAR12 = '';
$VAR13 = 'object_literal';
$VAR14 = 'Olympus';



$VAR1 = 'predicate_uri';
$VAR2 = 'http://www.someplace.org/rdf-image#format';
$VAR3 = 'subject_uri';
$VAR4 = 'http://www.gwmoore.org/ldip/ldip2201.jpg';
$VAR5 = 'predicate_ns';
$VAR6 = 'http://www.someplace.org/rdf-image#';
$VAR7 = 'object_lang';
$VAR8 = '';
$VAR9 = 'predicate_name';
$VAR10 = 'format';
$VAR11 = 'object_datatype';
$VAR12 = '';
$VAR13 = 'object_literal';
$VAR14 = 'jpeg';
```

The output is a parsable complete description of the triples contained in an RDF document (not shown). The RDF document was created as part of an image data specification project that is described in Chapter 18.

11.7. COMMON DATA ELEMENTS (CDES)

The term "common data element" is a misnomer. Most people, when they first encounter this term, infer that a common data element holds data. This inference, although reasonable, is not true. A common data element is the metadata that describes a datum in a data record. In XML parlance, a CDE is an XML tag. The thing that makes a descriptor "common" is its common usage by a scientific community. The way CDEs are intended to work is that a scientific community creates a list of CDEs that describe the kinds of data that their members use. The members of the community will all use the same CDEs (XML tags) to annotate their data files.

11.8. ISO-11179 SPECIFICATION FOR CDES

One of the most calamitous errors in any CDE project is to assume that everyone who reads a metadata tag will automatically understand its intended meaning. ISO-11179 is a standard way to define CDEs with the necessary information for understanding their meanings. The most popular CDEs in existence are the Dublin Core CDEs. These are a set of file descriptors that were prepared by a committee of librarians who convened in Dublin, Ohio. The Dublin Core includes basic information about electronic documents, such as, the title of the document, the name of the person who created the file, the date the file was created, the date the file was modified, and a short description of the file. These are the items a library software agent would retrieve if it were building an index of Internet documents. The world of informatics would be a better place if everyone who created an HTML, XML, or RDF file would remember to include the Dublin Core CDEs.

Dublin Core CDEs have been prepared to comply with the ISO-11179 specification. We learned something about ISO-11179 back in Chapter 10 Section 2. Every effort to create a data specification for a knowledge domain should begin by collecting the common data elements for the domain and annotating each element with the ISO-11179 CDE descriptors. The ISO-11179 descriptors for two of the Dublin Core CDEs (Title and Creator) are shown below:

From: *http://dublincore.org/documents/1999/07/02/dces/*

Title

Identifier: Title

Version: 1.1

Registration Authority: Dublin Core Metadata Initiative

Language: en

Obligation: Optional

Datatype: Character String

Maximum Occurrence: Unlimited

Definition: A name given to the resource.

Comment: Typically, a Title will be a name by which the resource is formally known.

Creator

Identifier: Creator

Version: 1.1

Registration Authority: Dublin Core Metadata Initiative

Language: en

Obligation: Optional

Datatype: Character String

Maximum Occurrence: Unlimited

Definition: An entity primarily responsible for making the content of the resource.

Comment: Examples of a Creator include a person, an organization, or a service. Typically, the name of a Creator should be used to indicate the entity.

In the next section, we will see that a list of ISO-11179-compliant CDEs contains all the information that is necessary to create an RDF schema (formal data dictionary in RDF syntax).

11.9. RDF SCHEMAS

An RDF schema is a dictionary file that lists metadata elements that can be used in RDF data documents. The official extended name for RDF schema is RDF Vocabulary Description Language. An RDF document draws its metadata terms from zero or more external RDF schemas.

When two documents draw their metadata terms from the same RDF schema, their statements (triples) can be easily merged.

RDF schemas typically consist of Classes and Properties. The easiest way to think about Classes and Properties is to equate Properties with the typical metadata tags used in RDF triples. Classes can be thought of as groupings for the subjects of RDF triples.

Consider the triple:

"Jules Berman" "blood glucose level" "85"

"Jules Berman" is the subject of the triple and is a member of a class. Let us say that the class that interests us is Patient. An RDF schema might specify that Patient is a class and that the Patient class is a subclass of Person. Jules Berman is an instance of the class Patient, which is a subclass of the class Person.

The property in the triple is "blood glucose level." An RDF schema would designate "blood glucose level" as a Property. Properties only apply to classes specified in the RDF Schema. The property "blood glucose level" may apply to the class "Laboratory Test" and may also apply to the class "Patient." In RDF jargon, the property "blood glucose level" has a domain of "Laboratory Test" and a domain of "Patient."

The data of the triple is "85." RDF Schema does not impose a datatype on the data value. If we wanted, we could assign a value of "bubble gum" to the "blood glucose level." In Section 11.10, we will learn that we can impose data constraints on a property's description. In the case of "blood glucose level" we might choose to limit values of the property to integers between 1 and 1000.

It might not be evident at this point, but all of the metadata for a knowledge domain and all of the ontological relationships for the subjects of a knowledge domain can be described in a simple RDF schema consisting of Classes and Properties.

Let us examine some of the properties of RDF schemas (List 11.9.1).

LIST 11.9.1. IMPORTANT FEATURES OF RDF SCHEMAS

1. RDF schemas are written in XML, but are completely unlike XML schemas.
2. RDF schemas contain formal declarations and descriptions of the classes and properties that are used in RDF documents.
3. RDF schemas, like all RDF documents, have no predetermined order or composition and consist of statements expressed as

triples. The subject of every triple in an RDF Schema will be either Class or Property.

4. Every RDF schema can be thought of as a child of the W3C RDF Schema that defines the "super" classes Resource, Class, and Property. All RDF schemas will refer to the document that defines RDF syntax and to the document that defines the W3C's top-level schema, and therefore will begin something like this:

```
<?xml version='1.0' encoding='ISO-8859-1'?>
<rdf:RDF
xmlns:rdf="http://www.w3.org/1999/02/22-rdf-syntax-ns#"
xmlns:rdfs="http://www.w3.org/2000/01/rdf-schema#">
```

5. An RDF schema may be the child of many RDF schemas. This means that a Class in an RDF schema may be the subclass of a Class element contained in another RDF schema.
6. RDF documents consist of triples (subject, metadata, value). RDF documents usually reference one or more RDF schemas to instantiate the subject of each triple (that is, to tell us which class in an RDF schema the subject is an instance of) and to provide subjects with class-appropriate metadata (properties).
7. Documents composed of triples whose components are defined by RDF schemas can be used to completely specify data objects within a knowledge domain.
8. By completely specifying data objects in a knowledge domain, RDF specifications achieve most of the functionality of data standards.

Let us see how an ISO-11179-compliant CDE can be converted into an RDF schema Class element.

Once we have the CDE (List 11.9.2), it is a straightforward job to create an equivalent RDF schema Class (List 11.9.3).

Let us examine the RDF definition for the Reagent Class. Classes are defined in RDF exclusively through their ancestral relation. To build a class in RDF schema, you announce that the element is a Class. You also provide a unique locator (such as a URL) or a unique, universally understood descriptor (more on this later) for the element, a description of the element, and the name of the father class of the element.

That's all there is to do for classes. You don't need to list the subclasses of the class because the subclasses will list the class as their

LIST 11.9.2. CDE FOR REAGENT

```
Class Label:Reagent
versionInfo (required): 0.1
Registration Authority: Association for Pathology Informatics
Obligation:optional
Maximum Occurrence: Unlimited
Datatype: Literal
comment: Histologic_stain_reagents,
   tissue_fixation_reagents, and
other chemicals  employed in the laboratory. For example:
distilled_water, ethanol, hematoxylin, aluminum_sulphate
subClassOf:Class
Contributor:Bill Moore
Date_of_contribution:05-30-2006
```

LIST 11.9.3. RDF CLASS ELEMENT CREATED FROM A CDE

```
<rdf:Class rdf:about="http://www.ldip.org/ldip_sch#Reagent">
<rdfs:label>Reagent</rdfs:label>
<rdfs:comment>
Histologic_stain_reagents, tissue_fixation_reagents, and other
chemicals  employed in the laboratory. For example:
   distilled_water,
ethanol,  hematoxylin, aluminum_sulphate
</rdfs:comment>
<rdfs:subClassOf
rdf:resource
   ="xmlns:rdfs="http://www.w3.org/2000/01/rdf-schema#Class"/>
</rdf:Class>
```

father in their own schema entry. You don't need to list the properties of the class because the properties will list the classes whose data they describe.

Do classes in RDF schema remind you of anything? The classes in an RDF schema comprise an ontology. An ontology is a list of classes and their relationships You can think of an ontology as the "classy" half of an RDF schema. Classes become most useful when they have Properties.

11.10. PROPERTIES (THE OTHER HALF OF THE RDF SCHEMA)

Creating Properties

A property is a metadata element that is used to describe the data assigned to one or more class objects. Here is the CDE for a property named "dateTime" (List 11.10.1).

LIST 11.10.1. CDE FOR "DATETIME" PROPERTY

```
Identifier:ldip:dateTime
Property Label:dateTime
versionInfo: 0.1
Registration Authority: Association for Pathology Informatics
Language:en
Obligation:optional
Maximum Occurrence: Unlimited
Datatype: /[\+\-]{1}[\d]{8}\.[\d]{6}Z[\+\-]{1}[\d]{4}/
comment: ISO 8601 format of date and time.
domain:Event
range: http://www.ldip.org/ldip_xsd.xsd#iso8601
Contributor:Bill Moore
Date_of_contribution:05-30-2006
```

Here is the same CDE expressed as a Property in RDF schema (List 11.10.2).

LIST 11.10.2. AN RDF SCHEMA DECLARATION FOR THE `dateTime` **PROPERTY**

```
<rdf:Property
   rdf:about="http://www.ldip.org/ldip_sch#dateTime">
<rdfs:label>dateTime</label>
<rdfs:comment>
The date and time at which an event occurs, in ISO8601
   format
</rdfs:comment>
<rdfs:domain
   rdf:resource="http://www.ldip.org/ldip_sch#Event"/>
<rdfs:range
   rdf:resource="http://www.ldip.org/ldip_xsd.xsd#iso8601"/>
</rdf:Property>
```

Let us look at the dateTime property. The first line announces that we will be declaring a Property. The second line tells us the name of the Property (dateTime) and its URL (the current RDF schema document). The third line provides the label by which we will refer to the Property. This might come in handy if we had different names for the property in different languages. The comment includes a definition for the element. The next line specifies the domain (class) for the property. The domain of a property is the class for which the property may be used. In this case, the domain for which the dateTime property applies is Event. This makes sense. If you need to describe an event, you would want to include the time that the event occurred. A property for a class serves as a property for all of the subclasses of the class (because all the subclass instances are members of the ancestor class). Every Property must have a domain (a class or classes for which the Property may be used) and a Range (a specified kind of data that is described by the Property). A property may have multiple classes in its domain. When a property has multiple classes in its domain, all the classes in the domain share the same property (obviously). This achieves some of the functionality of multiclass inheritance without needing to instantiate multiple classes under a single object. This is a subtle concept, and does not need to be mastered at this time. Suffice it to say that as you create your own RDF schemas, you should try to design your Properties to apply to multiple classes, and you should try to instantiate objects under a single class.

Specifying a Datatype from Within an RDF Schema Property Element

Let us continue to examine the `dateTime` property. Recall that a triple consists of a subject followed by metadata (the property element) followed by the data. The property element describes the data. The range of the property element tells us what kind of data is described. In RDF schemas, the range of a property is often "Literal," an element defined in the RDF syntax document that refers to any character string. You can see immediately that describing the range of a property as a character string does little to constrain or structure the expected values for a data element.

In the `dateTime` property, we want the range of the property to be data that conform to the ISO-8601 date/time format. How do we convey the datatype of the data/time element in RDF? One of the few deficiencies of RDF is that it provides no intrinsic datatyping facility. XML schema has the facility for datatyping the values described by an XML tag. So, for our property range, we provide a resource (URL) that specifies an element in an XML schema (`.xsd`) file that defines the datatype we need.

This somewhat convoluted strategy merits an example. The range for the `dateTime` property is a resource:

```
<rdfs:range
rdf:resource="http://www.ldip.org/ldip_xsd.xsd#iso8601"/>
```

The resource points us to an `xsd` file on the Web, and to a particular element within the `xsd` file, labeled iso8601. Let's pretend we visit the file and extract the iso8601 element (List 11.10.3).

LIST 11.10.3. AN `xsd` DATATYPE TO CONSTRAIN THE RANGE OF THE `dateTime` PROPERTY

```
<simpleType name='iso8601'>
<!-- values of a dataTime must contain                -->
<!-- a plus or minus sign occurring zero or one       -->
<!-- times followed by 8 digits                        -->
<!-- followed by a period                              -->
<!-- followed by 6 digits                              -->
<!-- followed by the letter Z, T, or a space        -->
<!-- followed by a plus or minus sign occurring       -->
<!-- zero or one time, followed by 4 digits           -->
<xsd:restriction base='string'>
<pattern value=''[\+\-]?[\d]{8}\.[\d]{6}[ZT
   ][\+\-]{1}[\d]{4}"/>
</xsd:restriction>
</simpleType>
```

The essence of the datatype is found in the pattern value line:

```
<pattern value=''[\+\-]?[\d]{8}\.[\d]{6}[ZT][\+\-]{1}[\d]{4}"/>
```

This line uses a Regular Expression to specify a pattern to which the element must conform.

The `.xsd` element definition imposes a datatype pattern on the value of the data described by the property. A validating software agent would check an RDF document to determine if the data described by a property conforms to the range of the property element defined by the element in the `.xsd` resource specified by the property range.

Whenever we have datatype constraints on the ranges of RDF schema properties, we will need to prepare an XSD file that contains elements for all the datatypes referred under the property ranges.

Creating the External XSD Document to Datatype Our Property Ranges

XSD datatype files are very easy to prepare. Basically, you just list your datatypes and provide descriptors. The following generic file contains samples of the kinds of datatypes you will probably need (patterns, inclusive values, and unions).

LIST 11.10.4. SAMPLE XML SCHEMA (.xsd FILE) FOR DATATYPING

```
<?xml version="1.0" encoding="UTF-8"?>
<xsd:schema
xmlns:xsd ="http://www.w3.org/2000/10/XMLSchema#">

<simpleType name='sp_pattern'>
<!-- values of an accession number must contain       -->
<!-- the letters sp followed by a hyphen followed     -->
<!-- by two digits followed by a hyphen                -->
<!-- followed by any number of digits                  -->
  <xsd:restriction base='string'>
    <pattern value='sp\-[0-9]{2}\-[0-9]+'/>
  </xsd:restriction>
</simpleType>


<xsd:simpleType name="adult_age">
<!-- an adult is at least 18 years old                 -->
  <xsd:restriction base="xsd:positiveInteger">
     <xsd:minInclusive value="18"/>
  </xsd:restriction>
</xsd:simpleType>

<xsd:simpleType name="serialNumber">
  <!--  may be either integers or mixed alphanumeric strings -->
  <xsd:union>
     <xsd:simpleType>
       <xsd:restriction base='integer'/>
     </xsd:simpleType>
     <xsd:simpleType>
       <xsd:restriction base='string'/>
     </xsd:simpleType>
  </xsd:union>
</xsd:simpleType>
```

```
<xsd:simpleType name="EnumerationObjectives">
<!--  may be either integers or mixed alphanumeric strings -->
  <xsd:restriction base="string">
    <xsd:enumeration value="2.5x"/>
    <xsd:enumeration value="6.3x"/>
    <xsd:enumeration value="20x"/>
    <xsd:enumeration value="40x"/>
    <xsd:enumeration value="100x"/>
  </xsd:restriction>
</xsd:simpleType>
</xsd:schema>
```

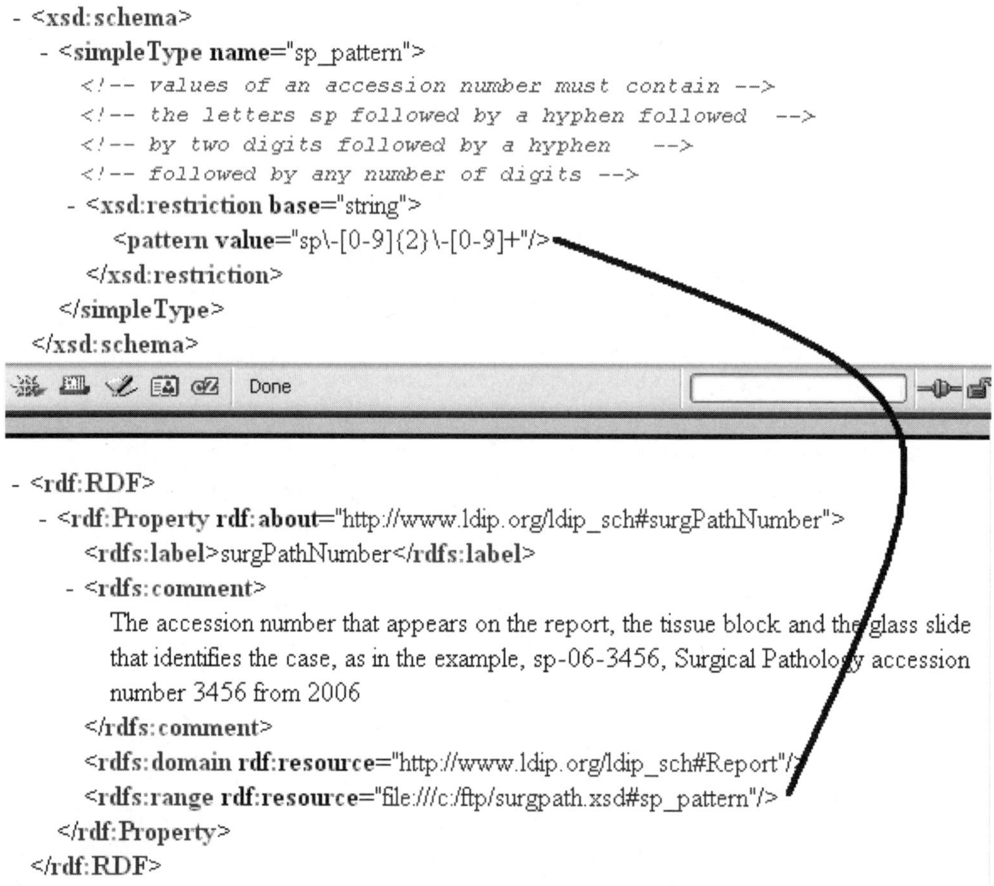

Figure 11-2 A Property's range constrains a Datatype through a link to an `xsd` document.

11.11. THE DIFFERENCES BETWEEN CLASSES AND PROPERTIES

The most difficult step in building any schema is determining whether a candidate element is a Class or a Property. Generalizations do not hold for all cases. For example, Classes tend to be nouns, while Properties (that describe data) tend to be adjectives. However, a Property can be a noun (for example, Time) if its role is to describe a data value (4:00 PM EST). Furthermore, we sometimes assign active processes to classes (for example, birth, death), and we cannot assume that classes are always static objects.

There is a strong tendency to assign subclass status to things that are not examples of their ancestral class. For instance, if Person is a class, someone may think that Leg is a subclass of Person (because a Leg is in a class of things that are parts of a Person). No! Leg is never a subclass of Person because a Leg is not a Person. A subclass of Person must be composed of types of Persons. So, Patient is a subclass of Person, and Pathologist is a subclass of Person, because they are both examples of Persons and because there are instances of Patients and instances of Pathologists. Remember, a class is a construct whose chief job is to provide specified instances. How about Friend? Is Friend a subclass of Person? Yes and no. Friend can be a subclass of person if you want to organize Persons based on whether they are Friends or not-Friends. However, if you think that being a friend is one of many features of any Person, you would be much better off defining friend as an RDF property. Recommendations for assigning Classes and Properties are found in List 11.11.1.

LIST 11.11.1. GENERAL RECOMMENDATIONS FOR DISTINGUISHING CLASSES AND PROPERTIES IN AN RDF SCHEMA

1. If something has instances of itself, it is almost always a class.
2. If a candidate class is a subclass of more than one class (so-called multiple inheritance), think very hard before making it a class. In most cases, you will be better off if the candidate is assigned as a Property or if it is excluded from the RDF schema.
3. Every class must be a subclass of a class. To be a subclass of a class the subclass must qualify as a member of the father class.
4. A class is fully specified when you know its definition and you know its ancestor class.
5. Properties describe data. If something has a specific datatype that includes numerics, it is almost always a property.

11.12. CREATING INSTANCES OF CLASSES

The purpose of a class is to support the creation of subclasses and class instances. If we have a `Report` class, we might also have a `Surgical_Pathology_Report` class which is a `subClassOf` Report. `Elsewhere_General_S06_4352` may be one unique instance of the class `Surgical_Pathology_Report`. As an instance of the class, the data in the report can be described using the properties specified in an RDF schema under the `Surgical_Pathology_Report` domain. The way to create an instance of a class in an RDF document is with the RDF "type" primitive. Here is an example of an RDF statement that tells us that a particular file ("*http://www.gwmoore.org/ldip/ldip2103.jpg*") is an instance of the class Image (List 11.12.1).

LIST 11.12.1. CREATING A CLASS INSTANCE WITH RDF "TYPE"

```
<rdf:description
   rdf:about="http://www.gwmoore.org/ldip/ldip2103.jpg">
<rdf:type resource= "http://www.ldip.org/ldip_sch#Image">
</rdf:description>
```

Whenever we want to create an instance of a class belonging to any chosen RDF schema, we will add an RDF statement much like the preceding one. We begin by specifying the subject of the statement (with the "about" declaration), then we indicate that the subject's "type" is a class listed in an RDF schema. An object may be an instance of more than one class, and a proper RDF statement may list numerous type/class pairs, but we caution that doing so adds complexity to your document.

11.13. PRESERVING NAMESPACES FOR CLASSES AND PROPERTIES

One of the most important features of RDF schemas is that you can mix and match different elements (classes and properties) from different schemas within a single document. This is done using a simple namespace notation that is common to all XML documents (List 11.13.1).

Notice that the elements camera and format appear twice, but on each occasion, they are prefixed by different namespaces (schem1 and schem2). The namespaces preserve metadata individuality.

Unique Subject Identifiers

Once you have made an instance of a class, you need to identify the instance uniquely. Failing this, the metadata-data pairs associated with a class instance have no specific meaning.

LIST 11.13.1. MIXING DIFFERENT RDF SCHEMA NAMESPACES IN A SINGLE RDF DOCUMENT

```
<?xml version='1.0' encoding='ISO-8859-1'?>
<rdf:RDF
xmlns:rdf="http://www.w3.org/1999/02/22-rdf-syntax-ns#"
xmlns:rdfs="http://www.w3.org/2000/01/rdf-schema#"
xmlns:dc="http://purl.org/dc/elements/1.1/#"
xmlns:scheml="http://www.someplace.org/#"
xmlns:schem2="http://www.someplace_else.com/#">


<rdf:Description
rdf:about="http://www.gwmoore.org/ldip/ldip2201.jpg">
<dc:creator>Bill Moore</dc:creator>
<scheml:camera>yes</scheml:camera>
<schem2:camera>Olympus</schem2:camera>
<scheml:format>jpeg</scheml:format>
<schem2:format>jpg</schem2:format>
</rdf:Description>
</rdf:RDF>
```

The typical unique subject identifier in an RDF triple is a URL specifying a unique Web location for a data object. Failing this, any unique identifier that permanently, unmistakably, and uniquely links an object to a character string will suffice. There are several registry services that provide identifiers for data objects in their domains. (Examples are shown in List 11.13.2.)

LIST 11.13.2. EXAMPLE OF REGISTRIES FOR OBJECT IDENTIFIERS

- **DOI**: Digital object identifier
- **PMID**: PubMed identification number
- **LSID**: Life Science Identifier
- **HL7 OID**: Health Level 7 Object Identifier
- **DICOM**: Digital Imaging and Communications in Medicine identifiers
- **ISSN**: International Standard Serial Numbers
- **SSN**: Social Security Numbers, for U.S. population
- **NPI**: National Provider Identifier, for physicians
- **PRS**: Clinical Trials Protocol Registration System

- **OHRP**: Office of Human Research Protections
- **FWA**: FederalWide Assurance number
- **DUNS**: Data Universal Numbering System
- **DNS**: Domain Name Service

In the life sciences, the LSID number has achieved some popularity. The LSID resolution protocol has five parts: Network Identifier (NID); root DNS, name of the issuing authority; namespace, chosen by the issuing authority; object ID, unique in that namespace and assigned locally; and revision ID, for storing versioning information (optional).

LSIDs can be used as URN's (Unique Resource Names) that uniquely identify items in RDF statements (List 11.13.3).

HL7 also provides unique identifiers. An enterprise can obtain an OID at: *http://www.iana.org/cgi-bin/enterprise.pl.*

For example, the University of Michigan OID is 1.3.6.1.4.1.250.

The enterprise OID serves as a prefix for unique data objects assigned by an institution.

Unique identifiers are used to uniquely specify the subject of a triple (that is, to specify what a triple is about) (List 11.13.4). Here we have a unique data object specified with an LSID for a PubMed citation. The number 8718907 is the unique PubMed citation number. We add a property/value pair consisting of the Dublin Core creator element and the data value, "Bill Moore." Once we have a unique subject, we can instantiate the element for

LIST 11.13.3. LSID EXAMPLES.

```
urn:lsid:pdb.org:1AFT:1
This is the first version of the 1AFT protein in the Protein
   Data Bank.

urn:lsid:ncbi.nlm.nih.gov:pubmed:12571434
This references a PubMed article.

urn:lsid:ncbi.nlm.nig.gov:GenBank:T48601:2
This refers to the second version of an entry in GenBank
```

LIST 11.13.4. USING A UNIQUE IDENTIFIER IN AN RDF "ABOUT" DECLARATION

```
<rdf:description
   rdf:about="urn:lsid:ncbi.nlm.nih.gov:pubmed:8718907">
<dc:creator>Bill Moore</dc:creator>
</rdf:description>
```

an appropriate class: `<rdf:descriptionrdf:about="urn:lsid:ncbi.nlm.nih.gov:pubmed:8718907">`

```
<rdf:type resource=
"http://www.ldip.org/ldip_sch#Document"/> <dc:creator>Bill
Moore</dc:creator>
</rdf:description>
```

Here we have a unique data object instantiated as a member of the Document class. The Document class is defined in an RDF schema referenced to a URL.

To summarize, the subject of a triple must be identified. The subject of a triple can be in the form of a URL (complete Web address) or a URN (Unique Resource Name). URLs and URNs are both forms of URIs (Unique Resource Identifiers).

You can create your own uniquely specified data object by appending a unique number to a URN prefix. For instance, a surgical pathology report, a patient name, or an image file can be the subject of a triple if it is identified by the following URN: `www.ldip.org:ldip:4Ib30fk6J3Y9gWpwMV27`.

Here, the prefix is "`urn:www.dlip.org:ldip:`". An alphanumeric suffix, "`4Ib30fk6J3Y9gWpwMV27`", is a 20-character random string that we chose for the object. There are many ways of providing identifiers for subjects. Once a subject is identified, triples containing the identifier can be merged from multiple RDF documents appearing anywhere on the Internet.

11.14. VALIDATING RDF

We will not discuss RDF validation at any length. Readers should be aware that validation is a necessary component of any RDF-based software applications. Minimal features of validating software are shown in the following lists.

LIST 11.14.1. CONDITIONS THAT VALIDATE A TRIPLE

1. The triple is expressed in proper RDF syntax.
2. The property in a triple is suitable for the subject.
3. The value of the triple is suitable for the property.

LIST 11.14.2. CONDITIONS THAT VALIDATE AN RDF DOCUMENT

1. The document is well-formed XML.
2. The document is well-formed RDF.
3. The triples are valid.

In the case of RDF that uses external XML schema documents to constrain the data type of the values included in triples, the software will also need to verify that the data of a triple conforms to the XML schema identified in the range of the property that describes the value.

11.15. RDF, SEMANTIC LOGIC, AND BIOMEDICAL ONTOLOGIES

Our ability to understand properties of data objects is always enhanced when the objects can be formally related to other objects in a knowledge domain. The term "pityriasis lichenoides et varioliformis acuta" might mean nothing to most people. If we are informed that this term is the name of a disease, and that the disease is a condition of the skin, we learn that the term shares the properties common to all diseases and, more specifically, to diseases involving the skin. Collections of defined objects and their relationships are called ontologies. RDF can be used to express relationships between objects, object properties, and object classes.

RDF was designed as a model of simplicity and logic. Unfortunately, soon after the release of the RDF specification, people started to notice certain limitations. As previously discussed, the most important limitation of RDF relates to datatyping. RDF simply does not contain a facile method for explicitly restricting the permissible values of a data element. In addition, RDF lacks certain concepts that ontologists would like to include, such as class union and class intersection.

DAML is an expansion of RDF that provides a variety of logical class operations that are considered essential in the field of artificial intelligence. The underlying assumption is that computed inferences of the kind needed in the design of software agents will require these logical constructs.

A more prosaic feature of DAML is its ability to reach into XML schema to assign metadata primitives that describe datatypes (List 11.15.1).

LIST 11.15.1. A FEW EXAMPLES OF XML SCHEMA PRIMITIVES THAT CAN BE INCORPORATED IN DAML

- enumeration
- positiveInteger
- minInclusive
- integer
- pattern

An example of the use of "positiveInteger" as specified within XML schema, and assigned within a DAML declaration, is:

```
<daml:DatatypeProperty rdf:ID="glucoseValue">
<rdfs:label>Glucose value</rdfs:label>
<rdfs:domain rdf:resource="#Test>
<rdfs:range rdf:resource="http://www.w3.org/2000/10/
     XMLSchema#positiveInteger">
</daml:DatatypeProperty>
```

From the Perl perspective, the most versatile XML schema primitive is "pattern," which permits DAML declarations to describe data ranges with regular expressions.

In the time-honored tradition of never leaving well enough alone, the DAML extension of RDF has itself been extended by OWL (Web Ontology Language). Like DAML, OWL provides a set of classes and properties that have a specific meaning (within the OWL schema). OWL users are guaranteed a common semantic interpretation of those triples that instantiate the OWL schema.

For the most part, the OWL specification includes classes and properties that permit logical inferences to be drawn from formal ontologies (52).

11.16. DATA SPECIFICATIONS CONTRASTED WITH DATA STANDARDS

An RDF schema is a dictionary of classes and properties. An RDF document consists of subject-metadata-data triples wherein subjects may belong to classes listed in RDF schemas and metadata may belong to properties listed in RDF schemas. By conveying scientific information in RDF documents, the functionality of standards are achieved minus the many limitations of standards. (List 11.16.1).

LIST 11.16.1. RDF FEATURES LACKING IN DATA STANDARDS

1. RDF data specifications are resources that anyone may choose to use or to ignore. The purpose of a data specification is to provide an opportunity for people to describe their data objects. Data standards, unlike data specifications, are often imposed requirements.
2. RDF data specifications are self-describing and contain all the information needed to interpret the contained information. Data files that conform to standards are often inscrutable and not intended to be read by humans.
3. General autonomous software agents can interrogate RDF data specifications. Competently written general software agents can parse and understand any RDF document. This is the underlying premise of the semantic Web. Files that conform to most data standards can be parsed only by software specifically written to accommodate the data standard.
4. RDF specifications can reduce the complexity of data. All data can be described in RDF documents as data triples. Data standards have no unifying principle of data description. The presence of competing standards, different versions of standards, and proprietary extensions of standards have contributed to the complexity of electronic information.
5. The data in a data specification can be distributed over multiple RDF documents.
6. The assertions in RDF data specifications have meaning, and the meaning is preserved when the assertion is extracted from the data-specification document. Data standards do not contain meaningful assertions. There is no universal way to extract components of a data standard and build datasets composed of meaningful assertions.
7. A data specification can comply with multiple RDF schemas at once.
8. A data specification can be written without violating intellectual property or breaching patient confidentiality.

11.17. SPECIFYING DATA WITH NOTATION 3

N3 (Notation 3) is a semantic alternative to RDF that preserves the meaning of data triples.

Consider the following free-text description of a pathology image:

"The image is a squamous cell carcinoma of the floor of the mouth. It was taken by Jules Berman on February 2, 2002. The

> microscope was an Olympus model 3453. The lens objective was 40x. The camera was a Sony model 342. The image dimensions are 524 by 429 pixels. The microscope and camera were not calibrated. The specimen is Baltimore Hospital Center S-3456-2001, specimen 2, block 3. The specimen was logged in 8/15/01 and processed using the standard protocol for H&E that was in place for that day. The patient is Sam Someone, medical identifier 4357. The tissue was received in formalin. The specimen shows a moderately differentiated, invasive squamous carcinoma. The patient has a 30-year history of oral tobacco use. The image is kept in `jpeg` (Joint Photographic Experts Group) file format and named `y49w3p2.jpg` and kept in the pathology subdirectory of the hospital's server. Its URL is *https://baltohosp.org/pathology/y49w3p2.jpg*. The image file has an md_5 hash value of 84027730gjsj350489. The image has no watermark. Copyright is held by Baltimore Hospital Center, and all rights are reserved."

Every assertion in the image description can be described by a triple consisting of a specified subject followed by a property followed by a value. Notation 3 (N3) is an abbreviated form of RDF. The narrative description can be readily translated in N3 (List 11.17.1).

LIST 11.17.1. IMAGE DESCRIPTION USING RDF TRIPLES IN NOTATION 3 FORMAT

```
file:image.n3 @prefix :
<http://www.pathologyinformatics.org/image_schema.rdf#>.
@prefix rdf: <http://www.w3.org/1999/02/22-rdf-syntax-ns#>.
:Baltimore_Hospital_Center rdf:type "Hospital".
:Baltimore_Hospital_Center_4357
    rdf:type"Unique_medical_identifier".
:Baltimore_Hospital_Center_4357 :patient_name "Sam_Someone".
:Baltimore_Hospital_Center_4357 :surgical_pathology_specimen
    "S3456_2001".
:S_3456_2001 rdf:type "Surgical_pathology_specimen".
:S_3456_2001 :image
    <https://baltohosp.org/pathology/y49w3p2.jpg>.
:S_3456_2001:log_in_date "2001-08-15".
:S_3456_2001 :clinical_history "30_years_oral_tobacco_use".
<https://baltohosp.org/pathology/y49w3p2.jpg> rdf:type
    "Medical_image".
<https://baltohosp.org/pathology/y49w3p2.jpg>
    :surgical_pathology_accession_number "S3456-2001".
<https://baltohosp.org/pathology/y49w3p2.jpg> :specimen "2".
```

```
<https://baltohosp.org/pathology/y49w3p2.jpg> :block "3".
<https://baltohosp.org/pathology/y49w3p2.jpg> :format "jpeg".
<https://baltohosp.org/pathology/y49w3p2.jpg> :width
   "524_pixels".
<https://baltohosp.org/pathology/y49w3p2.jpg> :height
   "429_pixels".
<https://baltohosp.org/pathology/y49w3p2.jpg> :hash_value
   "84027730gjsj350489".
<https://baltohosp.org/pathology/y49w3p2.jpg> :hash_type "md_5".
<https://baltohosp.org/pathology/y49w3p2.jpg> :watermark "none".
<https://baltohosp.org/pathology/y49w3p2.jpg> :camera "Sony".
<https://baltohosp.org/pathology/y49w3p2.jpg> :camera_model
   "342".
<https://baltohosp.org/pathology/y49w3p2.jpg> :capture_date
   "2002-02-02".
<https://baltohosp.org/pathology/y49w3p2.jpg> :diagnosis
   "squamous_cell_carcinoma".
<https://baltohosp.org/pathology/y49w3p2.jpg> :topography
   "floor_of_mouth".
<https://baltohosp.org/pathology/y49w3p2.jpg> :has
   "Intellectual_property_restriction".
<https://baltohosp.org/pathology/y49w3p2.jpg> :copyright
   "all_rights_reserved".
<https://baltohosp.org/pathology/y49w3p2.jpg>
   :copyright_holder "Baltimore_Hospital_Center".
<https://baltohosp.org/pathology/y49w3p2.jpg> :microscope
   "Olympus".
<https://baltohosp.org/pathology/y49w3p2.jpg>
   :microscope_model "3453".
<https://baltohosp.org/pathology/y49w3p2.jpg>
   :microscope_objective_power "40X".
<https://baltohosp.org/pathology/y49w3p2.jpg>
   :photographer_name "Jules_Berman".
```

This file (`image.n3`) can be converted to proper RDF with a Perl script (Lists 11.17.2, 11.17.3, and 11.17.4). The Perl script requires an external Perl module, `RDF::Notation3`, available from the `ActiveState ppm`.

We will revisit RDF and image annotation in Chapter 18, when we develop a suite of Perl scripts to build and implement an RDF schema for image annotations.

LIST 11.17.2. PERL SCRIPT `RDF_N3.pl`, A PERL SCRIPT THAT CONVERTS NOTATION 3 INTO RDF

```
#!/usr/local/bin/perl


#create the RDF equivalent of the N3 document
use RDF::Notation3::XML;
open(STDOUT,">image.xml");
$rdf = RDF::Notation3::XML->new();
$rdf->parse_file("image.n3");
$string = $rdf->get_string;
print $string;

exit;
```

LIST 11.17.3. PARTIAL OUTPUT OF `RDF_N3.pl`

```
<?xml version="1.0" encoding="utf-8"?>
<rdf:RDF
   xmlns:rdf="http://www.w3.org/1999/02/22-rdf-syntax-ns#">
<rdf:Description
   rdf:about="http://www.pathology
   informatics.org/image_schema.rdf#Baltimore_Hospital_Center">
<rdf:type
   xmlns:rdf="http
   ://www.w3.org/1999/02/22-rdf-syntax-ns#">Hospital</rdf:type>
  </rdf:Description>
<rdf:Description
   rdf:about="http://www.pathologyinfor
   matics.org/image_schema.rdf#Baltimore_Hospital_Center_4357">
<rdf:type
   xmlns:rdf="http://www.w3.org/199
   9/02/22-rdf-syntax-ns#">Unique_medical_identifier</rdf:type>
  </rdf:Description>
<rdf:Description
   rdf:about="http://www.pathologyinfor
   matics.org/image_schema.rdf#Baltimore_Hospital_Center_4357">
<patient_name
   xmlns="http://www.pathologyi
   nformatics.org/image_schema.rdf#">Sam_Someone</patient_name>
  </rdf:Description>
.
```

```
.
.
  </rdf:Description>
<rdf:Description
   rdf:about="https://baltohosp.org/pathology/y49w3p2.jpg">
<microscope_objective_power
   xmlns="http://www.pathologyinforma
   tics.org/image_schema.rdf#">40X</microscope_objective_power>
  </rdf:Description>
<rdf:Description
   rdf:about="https://baltohosp.org/pathology/y49w3p2.jpg">
<photographer_name
   xmlns="http://www.pathologyinforma
   tics.org/image_schema.rdf#">Jules_Berman</photographer_name>
  </rdf:Description>
</rdf:RDF>
```

LIST 11.17.4. PROPERTIES OF `image.n3`

- Every statement has a fully specified object followed by a property and a value (a triple). Every unique object belongs to a class (at least one).
- The N3 document can be interconverted to an equivalent RDF document.

11.18. RDF REDUCES COMPLEXITY

LIST 11.18.1. RDF REDUCES COMPLEXITY IN FOUR WAYS

1. RDF schemas are simple documents that may contain only classes and properties.
2. RDF schemas can be designed to avoid multi-class inheritance through the use of multi-class property domains. Avoiding multi-class inheritance reduces the complexity of drawing inferences from RDF documents.
3. The data in RDF documents exist in triples, and these triples preserve their meaning when extracted from the document in which they are contained.
4. A software agent that parses one RDF document should be able to parse any RDF document.

11.19. PERL METADATA MODULES

There are numerous Perl modules for handling metadata.

LIST 11.19.1. PERL METADATA MODULES

```
CORBA-XMLSchemas
Data-DumpXML
Libxml-perl
RDF-Core
RDF-Notation3
RDF-Simple
Spreadsheet-WriteExcel-FromXML
XML-Catalog
XML-Checker
XML-Clean
XML-Code
XML-CSV
XML-Database
XML-DOM
XML-Dumper
XML-Grove
XML-NamespaceSupport
XML-RSS-Aggregate
XML-RSS-Parser
XML-RSS
XML-SAX
XML-Schematron
XML-Schema
XML-Tidy
XML-XPath
XML-XQL
XML-XSLT
```

11.20. SUMMARY

RDF documents would not be useful if they could not be easily constructed, modified, transformed, merged, and analyzed by software agents. Much of the power of RDF comes from its generalizability. A software agent that interrogates one RDF document should be able to interrogate any RDF document. Once a generalized software agent is built, it becomes easy to make the agent autonomous (that is, able to traverse multiple RDF documents and to build inferences based on the meaningful assertions contained in those documents). RDF is likely to become the standard conveyance for biomedical data. It is important to all biomedical professionals to understand RDF and to have tools for handling RDF data. Perl provides simple access to RDF. We will revisit RDF in Chapter 18, when we implement a data-exchange specification.

CHAPTER

12

Mathematical Functions

12.1. BACKGROUND

Perl contains abundant math primitives, the basic functions upon which advanced mathematical algorithms are built. These include arithmetic functions, trigonometric functions, the modulus operator, a built-in pseudo-random number generator, exponential operations, and logical (bit-based) operators. Perl also supports bit-vector operations (that is, operations on arrays of bits).

The purpose of this chapter is to list some of the wide variety of mathematics resources available to Perl programmers and to demonstrate how biomedical professionals can employ mathematics to solve common tasks.

12.2. SIMPLE ADDITION

Adding numbers is somewhat trickier than one might imagine. Columns of numbers may contain mixed alphnumerics (p53, cd-117), or periods mixed with decimal points (He added 42.3 and 15. today.), as well as integers and floating point numbers. A simple Perl script that adds numbers found in a list file might vary depending on the expected content of the list and how the user intends to use the data.

LIST 12.2.1. PERL SCRIPT `add.pl` **ADDS COLUMN OF NUMBERS**

```
#/usr/local/bin/perl
open (TEXT, "add3.txt")||die"Can't open file";
$line = " ";
$count = 0;
$total = 0;
while ($line ne "")
    {
    $line = <TEXT>;
    if ($line =~ /[0-9]+\.[0-9]+/)
        {
        $number = $&;
        }
    elsif ($line =~ /\.[0-9]+/)
        {
        $number = $&;
        }
    elsif ($line =~ /[0-9]+/)
        {
        $number = $&;
        }
    else
        {
        next;
        }
    $total = $total + $number;
    print "$number $total\n";
    }
print "The total is $total";
exit;
```

The `add.pl` script produces some unexpected results with lists of numbers. What would the outcome be if the column of added numbers (in `add3.txt`) were as follows?

```
4
5
3.2222
0.4
3 .6
.6 3
```

he100llo

.

remember the 50th state

32

The output would be:

```
C:\ftp\pl>perl add.pl
4 4
5 9
3.2222 12.2222
0.4 12.6222
.6 13.2222
.6 13.8222
100 113.8222
50 163.8222
32 195.8222
The total is 195.8222
```

12.3. CUMULATIVE ADDITION OF A COLUMN OF NUMBERS

LIST 12.3.1. PERL SCRIPT `cum_add.pl` ADDS A COLUMN OF NUMBERS AND REPRESENTS EACH NUMBER AS A PERCENTAGE OF THE TOTAL

```
#/usr/local/bin/perl
open (TEXT, "add2.txt")||die"Can't open file";
$line = " ";
$count = 0;
$total = 0;
while ($line ne "")
    {
    $line = <TEXT>;
    $line =~ s/\n//o;
    $total = $total + $line;
    }
close TEXT;
open (TEXT, "add2.txt")||die"Can't open file";
$line = " ";
$count = 0;
$totalportion = 0;
```

```
while ($line ne "")
     {
     $line = <TEXT>;
     next if ($line eq "");
     $line =~ s/\n//o;
     $portion = $line / $total;
     $totalportion = $totalportion + $portion;
     print  "$line\|fract of tot $portion\|";
     print "cum fract of tot $totalportion\n";
     }
exit;
```

After each list entry is passed and summed, Perl calculates the "fract of tot," the fraction of the sum accounted for by the list element and the "cum fract of sum," the fraction of the sum accounted for by all of the list elements thus far encountered.

LIST 12.3.2. SAMPLE INPUT FILE FOR `cum_add.pl` SCRIPT

```
153.6
166.2
.
.
.
169.5
553.9
```

LIST 12.3.3. PARTIAL OUTPUT OF `cum_add.pl` SCRIPT

```
c:\ftp>perl cum_add.pl
153.6|fract of tot 0.0299695621634278|cum fract of tot
   0.0299695621634278
166.2|fract of tot 0.0324280028096464|cum fract of tot
   0.0623975649730742
.
.
.
169.5|fract of tot 0.0330718801217513|cum fract of tot
   0.891926168734879
```

12.4. POSIX

POSIX is the ISO/IEC 9945-1:1990 and IEEE Std 1003.1 Portable Operating System Interface. It is a large, standard library of programming operations developed with the National Institute of Standards, along with other organizations. About 250 functions are included in POSIX and most are mathematical. The purpose of POSIX is to provide a set of defined operations that will have the same name and functionality regardless of the language in which they are used. This provides significant interoperability between programming languages that adhere to the POSIX specification. Perl comes bundled with a POSIX module, providing direct access to these functions. To call a POSIX function, you simply invoke the POSIX module from your Perl script (Lists 12.4.1 and 12.4.2).

LIST 12.4.1. PERL SCRIPT `ceil.pl` CALLING TWO POSIX FUNCTIONS (CEIL AND FLOOR) FROM A PERL SCRIPT

```
#!/usr/local/bin/perl
use POSIX qw(ceil floor);
$num = 11.3;
print "Floor is ", floor($num), "\n";
print "Ceil is ", ceil($num), "\n";
exit;
```

LIST 12.4.2. OUTPUT OF `ceil.pl`

```
c:\ftp>perl ceil.pl
Floor is 11
Ceil is 12
```

In reality, it is difficult for programming languages to implement the full set of POSIX functions. The Perl interface to POSIX, which specifies functions that cannot be used in Perl, is available (53).

In addition, Perl users have developed and distributed a wide variety of advanced mathematical algorithms. Many of these algorithms, including Perl source code, can be found in *Mastering Algorithms with Perl*, by Jan Orwant et al (54).

12.5. A NO-FUSS COSINE WAVE

This program computes the cosine for all the degrees between zero and 90. Because these functions work on radians, the degree must be converted to radians. To do this, you multiply the degrees by pi over 180.

Perl doesn't have a built in pi value, but atan2(1,1) is pi divided by 45 degrees, so degrees are converted to radians with this formula.

LIST 12.5.1. PERL SCRIPT `coswav.p`, DISPLAYS 10 COSINE WAVE CYCLES ON THE SCREEN

```
#/usr/local/bin/perl
foreach (1...10)
  {
  for ($i=0;$i<360;)
  #foreach $degree (0..360)
     {
     $rads = (atan2(1,1) * $i) / 45;
     $length = abs(int(cos($rads) * 60)) ;
     print ("\|" x $length);
     print "\n";
     $i = $i + 6;
     }
  }
exit;
```

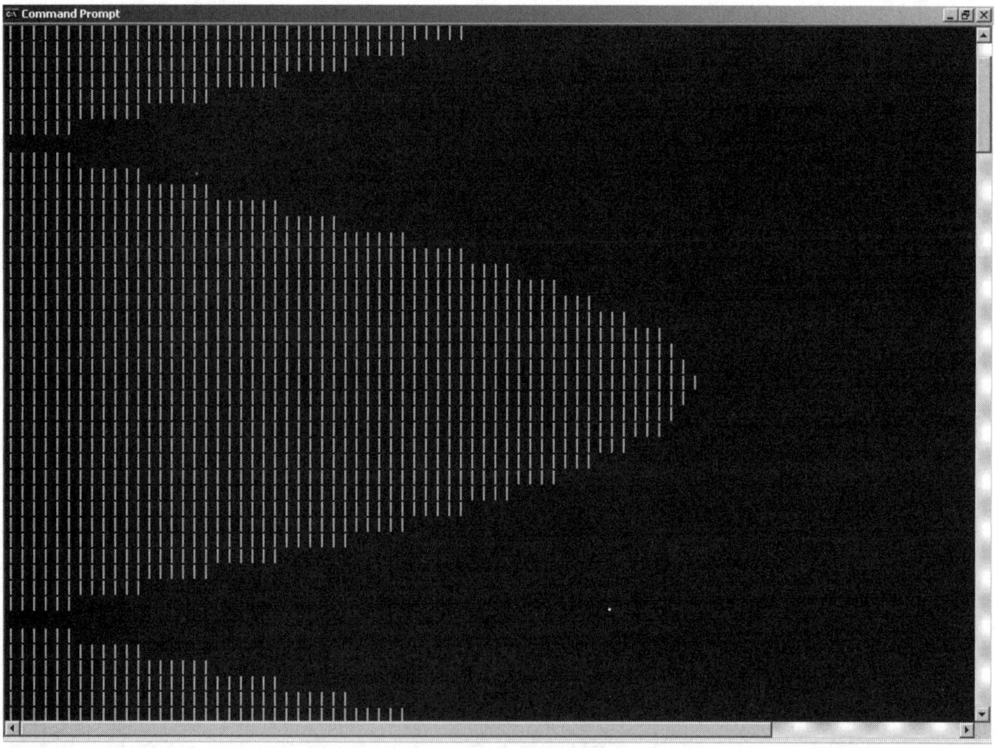

Figure 12-1 The output of `coswave.pl` is illustrated.

12.6. PERL MODULES FOR MATHEMATICS

Perl comes with built-in math operators, including addition, multiplication, modulo, exponents, logarithms, and trigonometric functions. There are also hundreds of external math modules that can be called from Perl scripts (List 12.6.1).

LIST 12.6.1. PERL MODULES FOR MATHEMATICS AND COMMONLY USED COMPUTER ALGORITHMS

```
Algorithm-ApplyDiffs
Algorithm-Combinatorics
Algorithm-CurveFit
Algorithm-Diff
Algorithm-Evolve
Algorithm-FastPermute
Algorithm-GenerateSequence
Algorithm-Hamming-Perl
Algorithm-Huffman
Algorithm-MarkovChain
Algorithm-PageRank
Algorithm-Pair-Best
Algorithm-Permute
Algorithm-Points-MinimumDistance
Algorithm-ScheduledPath
Algorithm-SixDegrees
Bit-Vector
Math-Calculus-Differentiate
Math-Calculus-Expression
Math-FFT (see Chapter 11)
Math-Matrix
Math-Random
Math::BigInt]
Math::Trig]
PDL (Perl Data Language)
POSIX (see Chapter 17)
BitstringSearch
Data-Search
Data-SearchReplace
Search-Binary
Search-FreeText
Search-VectorSpace
```

PDL (Perl Data Language) deserves special note. PDL is a library of math functions, many of which are devoted to matrix (specialized array) operations. PDL can be used in image processing and in graph plotting. Abundant information on PDL is available from: *http://pdl.perl.org* (55).

12.7. USING THE FAST FOURIER TRANSFORM MODULE

Jean Baptiste Joseph Fourier (1768-1830) gave his name to the Fourier transform. A transform is a mathematical operation that takes a function or a set of data and transforms it into something else. A reverse (or inverse) transform will take the product of the transform and produce the original function or dataset. For any transform, there might exist operations that are much easier to perform on the transformed function than on the original (untransformed) function.

Although the Fourier transform has applications in many areas of science, its value in biomedicine pertains mostly to signal processing (for example, images, ultrasound waveforms). The transform takes a time series representation of a signal (signal amplitude variation over time) and maps it into a frequency spectrum (occurrences at different frequencies). Periodicity in a signal is easy to find in a Fourier transform. Operations on a Fourier transform can eliminate periodic artifacts, eliminate frequencies that occur below a selected threshold, distinguish different signals (for example, voice and instrumental tracks), locate similarities between two signals, and so on. Although these activities are way beyond the scope of this book, it is heartening to know that the Fast Fourier Transform module (math-fft) can be easily called from the `ActiveState ppm`.

A brief and simple implementation of the math-fft is shown (List 12.7.1 and 12.7.2)

A medical image may be almost any kind of binary object that contains information related to a disease or physiologic process. Infrared spectroscopy, Doppler imaging, duplex scans, tissue microarray, whole slide scans, and multispectral analysis are just some examples of techniques that organize data into image objects when the native representation of the data is nonvisual or complex. Biomedical professionals are likely to use transforms (Fourier transforms, wavelet transforms, fractal transforms) and other mathematical algorithms (including vector quantization) to find patterns of data that signify the presence of pathognomonic cellular events. In Chapter 18, we will describe ways of annotating biomedical images regardless of their binary representations.

LIST 12.7.1. PERL SCRIPT `fft.pl` DEMONSTRATES THE FAST FOURIER TRANSFORM MODULE

```
#!/usr/local/bin/perl
use Math::FFT;
my $PI = 3.1415926539;
my $N = 8; #N can be any power of 2, such as 4,8,16,64
$series = [1,2,3,4,5,6,7,8,9,10,11,12,13,14,15,16];
#could be anything
print "series " . join(" ",@$series). "\n";
my $fft = new Math::FFT($series);
my $coeff = $fft->rdft();
print "coefficients \n @{$coeff}\n\n";
my $spectrum = $fft->spctrm;
print "spectrum \n @{$spectrum}\n";
exit;
```

LIST 12.7.2. OUTPUT OF FAST FOURIER TRANSFORM SCRIPT

```
C:\FTP>perl fft.pl
series 1 2 3 4 5 6 7 8 9 10 11 12 13 14 15 16

coefficients
136 -8 -7.99999999999999 -40.2187159370068 -8
-19.3137084989848 -8
-11.9728461013239 -8 -8 -8 -5.3454291
0335439 -8 -3.31370849898476 -8 -1.59129893903726

spectrum
72.25 13.1370711845441 3.41421356237309 1.61991440442178 1
0.723231346085845 0.585786437626905 0.5197830
6494829 0.25
```

CHAPTER 13

Statistics and Epidemiology

13.1. BACKGROUND

Most biomedical efforts will involve statistical tests. Fortunately, there are many excellent statistics packages available to researchers. The R statistical programming language has emerged as a popular and versatile tool in biomedicine (see Appendix). However, those who prefer to program their own methods in Perl will find that common statistical methods can be readily implemented.

Epidemiology draws heavily from statistics to yield clinically useful inferences on population data. Much of the epidemiological data available to biomedical professionals is made freely available by government agencies. In the United States, sources of epidemiologic data are the CDC, the National Cancer Institute's SEER (Surveillance, Epidemiology, and End Results) project, the U.S. Census, the U.S. Department of Vital Health Statistics, and the CMS (Centers for Medicare and Medicaid Services) public use data (see Appendix).

The purpose of this chapter is to describe a few techniques for acquiring and analyzing statistical data.

13.2. SIMPLE STATISTICS

The easiest and most fundamental statistical tests involve computing the mean of a population (List 13.2.1).

This simple program takes an array (provided in the script code) and prints the mean (average) of the array (sum of elements divided by the size of the array). Perl has several ways to determine the size of an array. The easiest is to force an array into scalar context using `scalar(@array`. This returns a string variable containing the number of elements in the

LIST 13.2.1. PERL SCRIPT `mean.pl` COMPUTES THE MEAN FROM AN ARRAY OF NUMBERS

```
#!/usr/bin/perl
@numbersarray = (1,2,3,4,5,6,7,8,9,10);
$arraysize = scalar(@numbersarray);
print "The number of elements in our array is $arraysize\n";
$sum = 0;
foreach $value(@numbersarray)
  {
  $sum = $sum + $value;
  }
$mean = $sum / $arraysize;
print "Your population number is $arraysize\n";
print "The array mean is $mean\n";
exit;
```

array. To determine the sum of all the elements in the array, we loop through the array, adding the value of each element to a summation variable. The mean is determined by dividing the sum of the elements by the number of elements, using the division operator,"/".

In practice, you will be computing the mean from files containing lists of numbers. Assuming your file consists of records containing numbers, you would build the array using an algorithm (List 13.2.2).

LIST 13.2.2. GENERAL METHOD OF BUILDING AN ARRAY THAT CAN BE USED IN A STATISTICAL OR MATHEMATICAL PERL ROUTINE

```
Open the file containing your records.
Go through the file, one line (record) at a time.
From a complex record, pick out the number you want using Regex.
Add that number to your array variable (using the Perl push
   command).
Calculate the mean (or any other statistical test) on the
   final array variable.
```

Or, you might want to collect numbers directly from the keyboard. The next script, `mean2.pl`, prompts you to start entering numbers, pressing the return key after each number. If you press the return key without entering a number, the script assumes you have finished your numbers and produces an output that contains the list, along with the computed mean (Lists 13.2.3 and 13.2.4).

LIST 13.2.3. PERL SCRIPT `mean2.pl` **COMPUTES THE MEAN OF A LIST ENTERED AT THE KEYBOARD**

```
#!/usr/bin/perl

#computes the mean of an array of numbers entered at keyboard
print "Type a bunch of numbers, pressing the return key\n";
print "after each number. Decimal numbers are allowed\n\n";
$number = " ";
until ($number eq "")
  {
  $number = <STDIN>;
   $number =~ s/\n//o;  #deletes the newline character
   if ($number eq "")
     {
     next;
     }
if ($number !~ /[0-9]+/)
   #the entry must contain at least one digit
     {
     print "You're only allowed to enter numbers...";
     print " We just won't count this entry\n";
     next;
}
if ($number !~ /^[0-9]*[\.]?[0-9]*$/)
#checks that the entry was a number
{

     print "You're only allowed to enter numbers...";
     print " We just won't count this entry\n";
     next;
     }
   push(@numbersarray, $number);
   }
$arraysize = scalar(@numbersarray);
$sum = 0;
foreach $value(@numbersarray)
  {
  $sum = $sum + $value;
  }
$mean = $sum / $arraysize;
print "@numbersarray\n";
print "Your population number is $arraysize\n";
print "The population mean is $mean\n";
exit;
```

LIST 13.2.4. OUTPUT OF `mean2.pl`

```
C:\ftp>perl mean2.pl
Type a bunch of numbers, pressing the return key after each
   number. Decimal numbers are allowed


11 12 13 11.111 12.33345 10 109hello12435
You're only allowed to enter numbers... We just won't count
   this entry 10


11 12 13 11.111 12.33345 10 10
Your population number is 7 The population mean is
   11.3492071428571
```

The `mean2.pl` script permits the user to enter numbers via the keyboard. After each entry, it determines:

1. Whether anything was actually entered (pressing the return key without entering anything is the signal to compute and display the mean).
2. Whether the entry was a valid number (zero or more digits separated by zero, or one decimal point followed by zero or more digits).

It is interesting to note that a simple decimal point unaccompanied by digits on either side meets criterion two for a valid number. We do not want this to happen, so we add the additional requirement that entries must contain at least one digit:

```
if ($number !~ /[0-9]+/)  #the entry must contain at
least one digit
```

13.3. COMPUTING THE STANDARD DEVIATION FROM AN ARRAY OF NUMBERS

The standard deviation is a commonly used quantifier of population variance from the mean. The following Perl script (List 13.3.1) is a modification of a program appearing in the superb resource book, *Mastering Algorithms with Perl* (54).

The sample script output is shown (List 13.3.2).

LIST 13.3.1. PERL SCRIPT `std_dev.pl` **COMPUTES THE STANDARD DEVIATION**

```
#!/usr/bin/perl
@numbersarray = (1,2,3,4,5,6,7,8,9,10);
#sample set of numbers
$arraysize = scalar(@numbersarray);
$sum = 0;
foreach my $value (@numbersarray)
  {
  $sum = $sum + $value;
  }
$samplemean = $sum / $arraysize;
#now, let us compute the square of the deviations from the mean
#and put these into an array
foreach my $value (@numbersarray)
  {
  $diffsquare = ($value - $samplemean) ** 2;
  push (@diffsquarearray, $diffsquare);
  }
#now, let us determine the mean of all the different squares
$sum = 0;
foreach my $value (@diffsquarearray)
  {
  $sum = $sum + $value;
  }
$diffsquaremean = $sum / ($arraysize - 1);
#finally, we take the square root of this mean, using Perl's
#built-in sqrt function
$std = sqrt ($diffsquaremean);
print "The mean is $samplemean\n";
print "The standard deviation is $std\n";
exit;
```

LIST 13.3.2. OUTPUT OF `std_dev.pl`

```
C:\ftp>perl std_dev.pl

The mean is 5.5
The standard deviation is 3.02765035409749
```

The standard deviation is the square root of the mean of the squares of the deviations from the mean for each element in the sample. It serves as an approximation of the deviations from the mean of the elements of a population, *n*, with degrees of freedom *n*-1. The calculation is straightforward: Compute the sample mean (as in the prior section). Loop through each element of the sample array, computing the square of the difference between the element and the mean. Add up all the squares of differences and compute the mean of the squares of differences. Finally, compute the square root of the mean of the squares of differences.

13.4. PERL MODULES FOR STATISTICS

Perl has many freely available statistics modules (List 13.4.1), all of which can be downloaded from CPAN or from the ActiveState Perl package manager (ppm).

Though it is possible to write simple Perl scripts for any statistical test, the author cautions against this practice. It is simply too easy to make a mistake when writing your own statistical scripts, and it is altogether too difficult to find script errors. It is much easier to use community-tested statistical modules that can be easily called from your scripts .

LIST 13.4.1. PERL MODULES FOR STATISTICS

Statistics-ChiSquare
Statistics-Contingency
Statistics-ConwayLife
Statistics-Descriptive
Statistics-Descriptive-Discrete
Statistics-Distributions
Statistics-Frequency
Statistics-GammaDistribution
Statistics-LineFit
Statistics-Lite
Statistics-RankCorrelation
Statistics-Regression
Statistics-ROC
Statistics-Shannon
Statistics-Simpson
Statistics-Table-F
Statistics-TTest

The standard distribution of Perl comes bundled with two great statistics modules: `Statistics::Basic` and `Statistics::Descriptive` (List 13.4.2).

LIST 13.4.2. METHODS IN THE `Statistics::Basic` MODULE

```
http://search.cpan.org/~jettero/Statistics-Basic-0.42/Basic.pm
Statistics::Basic::Mean;
Statistics::Basic::Median;
Statistics::Basic::Mode;
Statistics::Basic::Variance;
Statistics::Basic::StdDev;
Statistics::Basic::CoVariance;
Statistics::Basic::Correlation;
```

A sample script, `mean_var.pl`, computes the variance of a population using the `Statistics::Descriptive` module (List 13.4.3).

LIST 13.4.3. PERL SCRIPT `mean_var.pl` USES THE `Statistics::Descriptive` MODULE

```
#!/usr/local/bin/perl
use Statistics::Descriptive;
$stat = Statistics::Descriptive::Full->new();
$stat->add_data(1,2,3,4,5,6,7,8,9,10);
$mean = $stat->mean();
$var  = $stat->variance();
$std = $stat->standard_deviation();
print "mean $mean\n";
print "variance $var\n";
print "standard deviation $std\n";
exit;
```

The output of `mean_var.pl` is shown:

```
c:\ftp>perl mean.pl
mean 5.5
variance 9.16666666666667
standard deviation 3.02765035409749
```

Notice that the value of the standard deviation computed with the `Statistics::Descriptive` module is the same value that was computed with out home-made script that computed a standard deviation for the same population (List 13.3.2).

A sample script, `chi.pl`, uses the `Statistics::ChiSquare` module (List 13.4.4).

LIST 13.4.4. PERL SCRIPT `chi.pl` USES THE `Statistics::ChiSquare` MODULE

```
#!/usr/bin/perl
use Statistics::ChiSquare;
print chisquare([1, 9, 1, 15, 4, 7]), "\n";
print chisquare([20, 20, 20, 30, 20, 20, 30 ]), "\n";
exit;
```

The output of the `chi.pl` script is shown (List 13.4.5).

LIST 13.4.5. OUTPUT OF `chi.pl`

```
C:\ftp>perl chi.pl
There's a <1% chance that this data is random.
There's a >50% chance, and a <70% chance, that this data is
    random.
```

13.5. U.S. CENSUS STATISTICS

The U.S. Census Bureau publishes data files of its decennial census (see Appendix).

We will examine records from the 5 mbyte `alldata6.csv` file:

http://www.census.gov/popest/states/asrh/files/sc_est2004_alldata6.csv

This is an example of a comma-delimited data file. Comma-delimited data files usually contain a line or an external file (List 13.5.1) that serves as the key to the records.

LIST 13.5.1. KEY TO THE ALLDATA6 CENSUS FILE

```
SUM
    LEV,STATE,REGION,DIVISION,SEX,ORIGIN,RACE,AGE,CENSUS2000POP,
    ESTIMA
    TESBASE2000,POPESTIMATE2000,POPESTIMATE2001,POPESTIMATE2002,
    POPESTIMATE2003,POPESTIMATE2004
```

The first four records of the data are shown (List 13.5.2).

LIST 13.5.2. FIRST FOUR RECORDS FROM DATA FILE

```
040,01,3,06,1,1,01,0,19163,19164,19291,19157,18607,18511,18707
040,01,3,06,1,2,01,0,866,866,892,1019,1145,1266,1277
040,01,3,06,2,1,01,0,18034,18035,18169,18318,17736,17678,17721
040,01,3,06,2,2,01,0,869,869,895,951,1126,1202,1222
```

Once the key is studied, it is easy to parse the data file to collect the information you need.

The following script, `alldata.pl` (List 13.5.3), opens the census file and sums the data to determine the total number of people counted in the census.

LIST 13.5.3. PERL SCRIPT `alldata.pl` SUMS THE CENSUS DISTRICTS YIELDING THE TOTAL U.S. POPULATION

```
#!/usr/local/bin/perl
open (TEXT, "alldata6")||die"Can't open file";
$line = " ";
$total = 0;
while ($line ne "")
     {
     $count++;
     $line = <TEXT>;
     @lineitems = split(/\,/,$line);
     $lineval = $lineitems[8];
     #the 8th element is the population number for a group
     $total = $total + $lineval;
     }
close TEXT;
print "The total US population is $total\n";
exit;
```

It takes just a few seconds for Perl to parse the file and compute the U.S. population counted in the census (List 13.5.4).

LIST 13.5.4. OUTPUT OF `alldata.pl` FILE

```
C:\ftp\cgi>perl alldata.pl
The total US population is 281421906
```

There are many available U.S. Census files. They have many uses. A common, yet important task for epidemiologists is to prepare a listing of the number and percentage of the U.S. population for each decade of age. These numbers are used to adjust incidence statistics based on the proportion of people of the age groups at risk and are also used to adjust incidence findings for the current census with the findings of groups from earlier census reports. The U.S. population numbers, stratified by decade, can easily be computed with Perl scripts once the comma-delimited data sets are acquired and the data elements are identified with the data record key.

13.6. SEER STATISTICS

For cancer epidemiologists, one of the most important resources is the SEER public-use data files (see Appendix). SEER collects cancer records from multiple geographic sites and distributes files as byte-designated flat files (List 13.6.1).

What is a byte-designated flat file? Byte-designated files are simple ASCII files in which each line holds a data record and every data record has the same length. A data dictionary provides the key to the byte-delineated data elements. Let us inspect a few records from a SEER file (List 13.6.2).

LIST 13.6.1. SOME OF THE ORGAN-SPECIFIC INFORMATION PUBLICLY AVAILABLE FROM SEER FILES

```
2002      34,785,520 BREAST.TXT
2002      23,085,062 COLRECT.TXT
2002      15,104,782 DIGOTHR.TXT
2002      20,239,758 FEMGEN.TXT
2002      16,321,074 LYMYLEUK.TXT
2002      32,392,052 MALEGEN.TXT
2002      28,956,158 OTHER.TXT
2002      28,339,542 RESPIR.TXT
2002      12,987,282 URINARY.TXT
```

LIST 13.6.2. FIRST FOUR RECORDS (TRUNCATED) OF SEER FILE "URINARY.TXT"

```
0200003549011009007191507702025000041993C67408120331999209980 0...
0200002232202100300719230720101202091995C659281203419993009800...
0200002756101100100719400540202100121994C679080103319999999800...
0200034355011001410192806701012001219 95C649283123210506099800...
```

A separate file distributed by SEER explains how each record is organized. Bytes 45 to 49 (List 13.6.3) contain the ICD code (International Classification of Disease) code for a neoplasm.

LIST 13.6.3. CHARACTERS 45-49 OF FOUR INDIVIDUAL SEER RECORDS

```
Characters 45-49 of record 1 of urinary.txt 81203
Characters 45-49 of record 2 of urinary.txt 81203
Characters 45-49 of record 3 of urinary.txt 80103
Characters 45-49 of record 4 of urinary.txt 83123
```

The codes 81203, 80103, and 83123 are the final five digits of an ICD oncology code (List 13.6.4). We have shown that we can extract the ICD oncology codes from the UMLS Metathesaurus MRCON file (Chapter 5 Section 5).

LIST 13.6.4. A FEW ICD RECORDS

```
C0007117|.....ICD10AM|PT|M8090/3|Basal cell carcinoma NOS|3|N||
C0007118|.....ICD10AM|PT|M8094/3|Basosquamous carcinoma|3|N||
C0007118|.....ICD10AM|PT|M8095/3|Metatypical carcinoma|3|N|256|
C0007120|.....ICD10AM|PT|M8250/3|Bronchiolo-alveolar
   adenocarcinoma|3|N||
C0007120|.....ICD10AM|PT|M8251/3|Alveolar adenocarcinoma|3|N||
C0007124|.....ICD10AM|PT|M8500/2|Intraductal carcinoma,
   noninfiltrating NOS|3|N||
```

We can use the ICD nomenclature to construct the names of neoplasms (List 13.6.5) from bytes 45-49 of SEER records.

It is easy to prepare a Perl script that can parse through all of the SEER files, extracting and evaluating any information that we choose (Lists 13.6.6 and 13.6.7).

LIST 13.6.5. ICD NAMES FOR CODES CONTAINED IN CHARACTERS 45-49 OF FOUR SEER RECORDS

```
Record 1 of urinary.txt 81203 - Transitional cell carcinoma NOS
Record 2 of urinary.txt 81203 - Transitional cell carcinoma NOS
Record 3 of urinary.txt 80103 - Carcinoma NOS
Record 4 of urinary.txt 83123 - Renal cell carcinoma
```

LIST 13.6.6. PERL SCRIPT `SEER.pl` **FOR DETERMINING THE FREQUENCY OF OCCURRENCE IN THE UNITED STATES OF TUMOR TYPES FOUND IN THE SEER PUBLIC-USE DATA FILES**

```
#!/usr/local/bin/perl
open(TEXT,"ICD.TXT");
$line = " ";
while ($line ne "")
  {
  $line = <TEXT>;
  @linearray = split(/\|/,$line);
  $icdcode = $linearray[13];
  next unless ($icdcode =~ /^M[0-9]{4}\/[0-9]{1}/);
  $icdcode =~ s/^M([0-9]{4})\/([0-9]{1})/$1$2/;
  $codehash{$icdcode} = $linearray[14];
  }


@filelist = glob("C\:\\ftp\\seer\\*.TXT"); #see Glob in Glossary
open (OUT, ">seer.out");
foreach $value (@filelist)
  {
  open (TEXT, $value);
  $line = " ";
  while ($line ne "")
    {
    $count++;
    $line = <TEXT>;
    $code = substr($line,44,5);
    $code = "$code\|$codehash{$code}";
    $subhash{$code}++;
    }
  close TEXT;
  }


while ((my $key, my $value) = each(%subhash))
   {
   $fraction_of_total = $value / $count;
   $fraction_of_total = substr($fraction_of_total, 0, 6);
($fraction_of_total = "0.0000") if ($fraction_of_total =~
   /^[^0]/);
   $value = substr("00000$value",-6,6);
   $total = $total + $value;
```

```
        push(@result, "$value\|$fraction_of_total\|$key");
        }

    @result = reverse (sort(@result));
    foreach $thing (@result)
        {
        $rank++;
        $rank = substr("0000$rank",-4,4);
        print $rank . "\|$thing\n";
        print OUT $rank . "\|$thing\n";
        }
    exit;
```

The second number in each line of List 13.6.7 is the number of occurrences of the cancer specified in the final element of the record.

LIST 13.6.7. FIRST 20 LINES OF OUTPUT OF PERL SCRIPT `SEER.pl`

```
0001|401947|0.2916|81403|Adenocarcinoma NOS
0002|133958|0.0972|85003|Infiltrating duct carcinoma
0003|078938|0.0572|80703|Squamous cell carcinoma NOS
0004|064993|0.0471|80103|Carcinoma NOS
0005|038011|0.0275|81303|Papillary transitional cell carcinoma
0006|031461|0.0228|80772|Intraepithelial neoplasia, grade
   III, of cervix, vulva, and vagina
0007|023821|0.0172|80003|Neoplasm, malignant
0008|018961|0.0137|80413|Small cell carcinoma NOS
0009|018572|0.0134|87203|Malignant melanoma NOS
0010|018433|0.0133|84803|Mucinous adenocarcinoma
0011|017729|0.0128|85002|Intraductal carcinoma,
   noninfiltrating NOS
0012|017556|0.0127|83123|Renal cell carcinoma
0013|016681|0.0121|87433|Superficial spreading melanoma
0014|015785|0.0114|81203|Transitional cell carcinoma NOS
0015|015583|0.0113|85203|Lobular carcinoma NOS
0016|013534|0.0098|97323|Multiple myeloma
0017|013532|0.0098|96803|Malignant lymphoma, large cell,
   diffuse NOS
0018|012910|0.0093|80702|Squamous cell carcinoma in situ NOS
0019|012825|0.0093|83803|Endometrioid carcinoma
0020|011423|0.0082|80123|Large cell carcinoma NOS
```

Many people hate byte-designated data files, finding them inscrutable and generally useless. Actually, byte-designated data files are a simple, condensed way of conveying well-organized, large datasets. With a little patience, it is possible to understand their organization and fathom their content. The SEER files, in particular, contain a wealth of information related to the cancers that occur in the U.S. population. The information in SEER files can be integrated with other data sets by clever data miners.

13.7. RECEIVER OPERATOR CHARACTERISTIC (ROC) CURVES

ROC is the receiver operator characteristic (also known as the receiver operator curve). ROC was developed in World War II to help receiver operators (the people who watched radar screens) distinguish between enemy aircraft and blips due to noise. ROCs are now a popular method to determine a good threshold (sometimes called cut point) for a "positive" score on a diagnostic test (56).

Before providing a mathematical definition of a ROC, let us consider a prototypical instance where a diagnostic test yields accurate results, but cannot usefully determine whether a patient actually has the disease that is being evaluated.

In Figure 13-1, we see the fictitious results of PSA (prostate specific antigen, a blood marker often raised in patients with prostate cancer) testing on two populations of men. The population on the left represents the spread of PSA values in men who do not have prostate cancer (as tested by manual examination of the prostate, magnetic resonance imaging, and multiple prostate biopsy samplings of the prostate). The population on the right consists of men known to have prostate cancer, and their PSA scores. Let us assume that the test itself is a highly accurate measurement of PSA. An accurate and repeatable test is one in which the

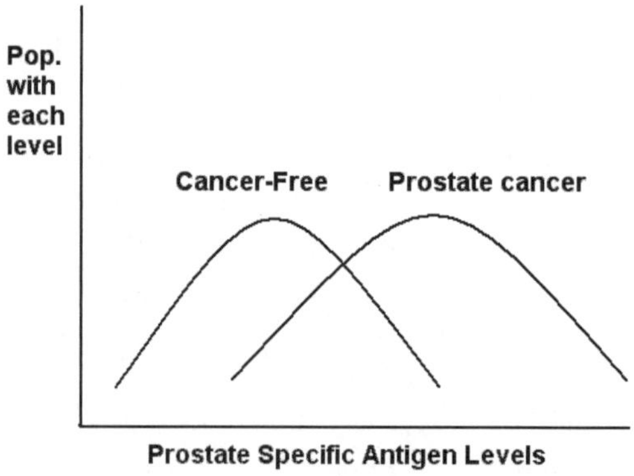

Figure 13-1 Dilemma of choosing a diagnostic cut-point to distinguish two populations.

measurement (in this case, PSA level) accurately reflects the level of PSA in the patient's blood, and that repeated measurements of PSA on the same sample of blood will yield very nearly the same result.

If we move up the scale of PSA values measured in either population of men, we will find that each value occurs in some men with cancer and in some men without cancer. We want to choose a cut-off value of PSA that detects men who are likely to have prostate cancer, but does not include many men who are not likely to have prostate cancer. If we pick a low PSA value as our cut-off value, we will catch most of the population of men who have cancer, but we will also catch a great number of men who do not have cancer. If we pick a high PSA value as our cut-off, we are likely to exclude most men who are cancer-free, but we will miss many of the men who have cancer. How do we choose the best value of PSA for our cut-off?

The ROC is a plot of the fraction of true positives (y axis) against the fraction of false positives (x axis) for all the possible measurements of a test result. ROC plots all look roughly the same, beginning at 0,0, ending at 1,1, and curving higher than the line for chance accuracy (a straight line at a 45-degree angle connecting 0,0 and 1,1). If the experimental ROC plot lies close to a 45-degree angle, the test is roughly equivalent to the accuracy of blind chance. If the text produces a curve that rises sharply upward, it is likely to be a good test. The cut-point for a good test depends on factors related to the medical tolerance for false positives or false negatives. In general, though, the cut-point will be somewhere near the upper-left-most point in the ROC plot.

A Perl module for computing ROC plots is available from the ActiveState Perl package manager and is described in a separate publication (57).

An example Perl script, `roc.pl`, using the `Statistics::ROC` module is shown here (List 13.7.1).

LIST 13.7.1. PERL SCRIPT `roc.pl` DEMONSTRATES `Statistics::ROC`

```
#!usr/bin/perl
use Statistics::ROC;
use Data::Dumper;
@var_grp =
   ([1.5,0],[1.4,0],[1.4,0],[1.3,0],[1.2,0],[1,0],[0.8,0],
          [1.1,1],[1,1],[1,1],[0.9,1],[0.7,1],[0.7,1],[0.6,1]);

   @curves=roc('decrease',0.95,@var_grp);
   print Dumper($curves[1]);
exit;
```

The roc.pl script receives three arguments:

1. A 'decrease' or 'increase', indicating that a lower ('decrease') value is a negative test result or a higher 'increase' is a positive result.
2. A two-sided confidence interval (default is 0.95).
3. The data provided is a list-of-lists, with each element value followed by a Boolean. The 0 stands for disease not present, and 1 stands for disease present.

The output of the roc.pl Perl script is a set of doublets representing the x, y coordinates of a ROC plot (List 13.7.2). In the sample script, only a few data values were plotted, producing a rough-angled plot. Typical ROC plots use many data points and produce a smooth, bow-shaped curve.

LIST 13.7.2. X,Y OUTPUT PAIRS FROM roc.pl

```
0,
0


0,
'0.0942763357317147'


0,
'0.29624309072361'


0,
'0.29624309072361'


'0.0942763357317147',
'0.398121545357938'


'0.161383149363309',
'0.499999999992266'


'0.228489962994904',
```

```
'0.703756909276412'


'0.228489962994904',
'0.703756909276412'


'0.228489962994904',
'0.703756909276412'


'0.29630302471766',
'0.905723664361806'


'0.364116086440415',
1


'0.499999999992266',
1


'0.703756909276412',
1


'0.703756909276412',
1


'0.905723664361806',
1


1,
1
```

The x,y coordinates can be easily imported into graphics software to compute a graph representing the ROC plot (see Figure 13-2). For this project, I used the free, open-source tool, Open Office 2.0 Calc and Draw (*http://www.openoffice.org*).

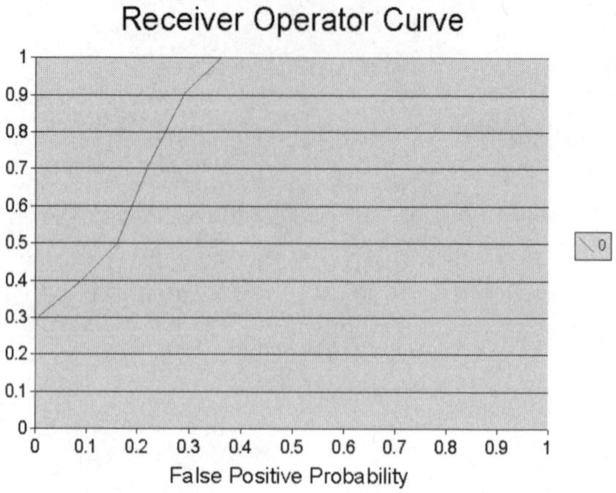

Figure 13-2 ROC plot of output data from `roc.pl`.

CHAPTER

14

Modeling Biological and Medical Systems

14.1. BACKGROUND

Mathematical models are important because they provide us some sense of the realm of possible outcomes of a biological experiment or a clinical trial. If we can anticipate the kinds of outcomes that might occur in a study, we might improve our trial designs to maximize the chance for a decisive outcome.

The following "thought" experiment demonstrates how a prospective clinical trial might fail to correctly answer a hypothesis.

In a retrospective study, Dr. X demonstrated a "perfect" tumor marker that never failed to distinguish between two tumor variants (aggressive and indolent). The aggressive tumor variant grew 10 times as fast and metastasized at 10 times the rate as the indolent variant. Without Dr. X's tumor marker, the two tumor variants could not be distinguished by any morphologic evaluation (that is, indolent and aggressive tumors looked alike).

Dr. X's perfect tumor marker was received with great enthusiasm. For the first time, it looked as though patients would be treated in a manner appropriate to the biology of their tumors. Patients with aggressive tumors would have resection followed by intensive chemotherapy. Patients with indolent tumors would have resection and no chemotherapy.

Of course, Dr. X would need to wait until a prospective clinical trial confirmed the clinical validity of his marker.

In a prospective trial of the same marker, 200 tumors are excised at the time of clinical detection (tumor size 2 cm) and receive no additional treatment. Dr. X finds that 100 of the tumors stain as "indolent variants" and 100 tumors stain as "aggressive variants." The prospective trial followed all 200 patients, determining survival at five years.

Here's the rub. At the end of the trial, there is no survival difference between patients with "indolent variants" and patients with "aggressive variants." The marker is considered a total failure, with millions of dollars wasted on the prospective trial.

How is this possible? Poorly designed prospective trials can lead to erroneous conclusions. The key to finding the error in this imaginary example lies in understanding why cancer patients might die after their tumor has been excised. Cancer deaths following primary tumor excision are almost always due to metastases that were seeded prior to the time the primary tumor was excised. If the primary tumor had not metastasized before it was excised, the patient is cured by the excision. If the primary tumor had metastasized before it was excised, the patients is likely to die. This is true whether the tumor is an aggressive tumor or an indolent tumor. The survival advantage of indolent tumors (compared to aggressive tumors) is that indolent tumors are less likely to metastasize.

Consider the design of the prospective trial and the initial conditions of the trial. Two hundred tumors were excised. All excised tumors (whether indolent or aggressive) were 2 cm. in diameter. We are told that the aggressive tumors have a growth rate 10 times that of the indolent tumors. So, if it took 1 month for an aggressive tumor to reach a 2-cm. size, it must have taken 10 months for an indolent tumor to reach 2-cm. size. We are also told that the rate of metastasis in the aggressive tumor is 10 times that of the indolent tumor. Because aggressive tumors had 1/10th the growth history of the indolent tumors (for instance, 1 month compared to 10 months), both the aggressive and indolent tumors had the same number of metastastic cases at the time that the tumors were excised (that is, at the moment the prospective clinical trial began). Hence, there was no difference in the survival outcome between the tumor variants, because survival outcome is almost entirely determined by the presence or absence of metastases at the time of primary tumor excision. A computational model that simulated the growth and metastases of the two tumor populations may have predicted the disappointing and misleading outcome of the clinical trial.

The purpose of this chapter is to provide a few modeling methods, implemented in Perl, that can be used by experimentalists and trialists to improve the design of their studies.

14.2. USING RANDOM NUMBERS

Many simulation programs rely on a random number generator. The random numbers are used to simulate probabilistic events.

Let us imagine that you are not exactly a whiz at probability. You have a pair of dice and want to know how often you might expect each of the numbers (from one to six) to appear after you have thrown one die (List 14.2.1).

LIST 14.2.1. PERL SCRIPT `randtest.pl` **SIMULATES 600,000 CASTS OF THE DIE**

```
#!/usr/bin/perl
$count = 0;
while ($count < 600000)
   {
   $count++;
   $one_of_six = (int(rand(6))+1);
   $hash{$one_of_six}++;
   }
  while(($key, $value) = each (%hash))
   {
   print "$key => $value\n";
   }
exit;
```

The essence of the Perl script is in the following line, which uses Perl's pseudo-random number generator, `rand()`:

```
$one_of_six = (int(rand(6))+1);
```

This command will randomly assign a number between 1 and 6 to the variable *$one_of_six*.

How does it work? `Rand(EXPRESSION)` returns a random fractional number greater than or equal to 0 and less than the value of *EXPRESSION*. If *EXPRESSION* is omitted, the value 1 is used.

We are not interested in a decimal fraction, so we use the integer function, `int()`, on the output of the `rand()` function to produce a 0, 1, 2, 3, 4, or 5. Because we want our output to be an integer that is 1, 2, 3, 4, 5, or 6, we simply add 1 to the output number.

Randtest starts by setting a loop, simulating 600,000 casts of the die. Each loop uses the rand function. The Perl function `rand(number)` yields a pseudo-random number of value less than 6.

We make a hash of the frequency of occurrence of the different integer outcomes for all of the 600,000 simulations.

The results are as one might expect. Each number "came up" about 100,000 times. Just for fun, we repeated the script output, with much the same result (Lists 14.2.2 and 14.2.3).

There are many uses of random numbers, particularly in the fields of cryptography and probability.

LIST 14.2.2. OUTPUT OF FIRST TEST OF `randtest.pl`

```
C:\ftp>perl randtest.pl
1 => 100002
2 => 99902
3 => 99997
4 => 100103
5 => 99926
6 => 100070
```

LIST 14.2.3. OUTPUT OF SECOND TEST OF `randtest.pl`

```
C:\ftp>perl randtest.pl
1 => 100766
2 => 99515
3 => 100157
4 => 99570
5 => 100092
6 => 99900
```

14.3. RESAMPLING AND MONTE CARLO SIMULATIONS

Random numbers are used in many artificial life programs. These programs are based on simulating objects that obey certain rules of behavior (for example, move one square in any direction, duplicate yourself, disappear) and events that are triggered by probabilistic outcomes occurring with timed moments or events.

In such simulations, there are many outcomes that could result from a small set of initial conditions. It is much easier to write these programs and observe their outcomes than to directly calculate outcomes from a set of governing equations.

Biological simulations can be modeled with a few lines of Perl and can be used to predict outcomes that would be intractable to any direct mathematical analysis (58), (59), (60), (61) (Lists 14.3.1 and 14.3.2).

The `ai.pl` script produces 10 simulations and outputs the results of each simulation. Each simulation begins with a single cell and allows the cell and any progeny cells to advance through division cycles. Each cell has a chance of dying during each division cycle. When a cell dies, the population of the colony is reduced by one. When a cell survives, the population of the colony is increased by two. After each division

LIST 14.3.1. PERL SCRIPT `ai.pl` **SIMULATES CLONAL TUMOR GROWTH**

```
#!/usr/bin/perl
print "Enter the death probability for your simulation\n";
print "Number must be between zero and one.\n";
print "Most realistic numbers are .45 to .50\n";
$value = <STDIN>;
$value =~ s/\n//o;
if ($value > 1)
  {
print "Exiting... you must pick a number between zero and
   one\n";
  end;
  }
print "THE CELL DEATH PROBABILITY FOR THIS SIMULATION IS
   $value\n\n";
my $roundnumber = 1; #initiate the generation counter
&cycle;


sub cycle
{
my $total = 0;
my $count = 1;
if ($roundnumber > 10)
  {
  print "I've seen enough!";
  end;
  }
$roundnumber++;
print "\nStarting with a single malignant cell,";
print " let us watch the clonal growth.\n";
$sum = 1;
while($sum > 0)
   {
   my $i = 1;
   while ($i < $sum +1)
     {
     $i++;
     $randnum = int( rand(100) ) + 1;
     if ($randnum > (100 * $value))
        {
        $sum = $sum + 1;
```

```
            }
        if ($randnum < ((100 * $value) -1))
            {
            $sum = $sum - 1;
            }
        }
    if ($sum < 1)
        {
        print "Tumor terminated...good!\n";
        &cycle;
        }
    if ($sum > 5000)
        {
print "\nBad news. Let us stop watching this malignancy.\n";
        &cycle;
        }
    if ($sum > 0)
        {
        if (length($total) < 60)
          {
          print "$sum ";
          $total = $total . $sum;
          }
        else
          {
          print "\n$sum ";
          $total = "$sum ";
          }
        }
    }
return 1;
}
exit;
```

LIST 14.3.2. OUTPUT OF `ai.pl`

```
C:\ftp>perl ai.pl
Enter the death probability for your simulation.
Number must be between zero and one.
Most realistic numbers are .45 to .50
.46
THE CELL DEATH PROBABILITY FOR THIS SIMULATION IS .46.
```

```
Starting with a single malignant cell, let's watch the
   clonal growth. Tumor terminated...good!
Starting with a single malignant cell, let's watch the
   clonal growth. 1 Tumor terminated...good!
Starting with a single malignant cell, let's watch the
   clonal growth. 2 1 4 2 1 1 1 Tumor terminated...good!
Starting with a single malignant cell, let's watch the
   clonal growth. 2 1 5 6 8 8 8 12 15 18 18 20 19 27 32 30 31
   20 14 16 23 30 30 36 38 34 50 52 67 75 97 114 133 143 150
   156 159 178 200 254 302 292 329 336 382 441 489 603 630 701
   770 862 923 1056 1084 1210 1369 1473 1664 1776 1959 2196
   2475 2862 3098 3327 3740 4095 4634 Bad news. Let's stop
   watching this malignancy.
Starting with a single malignant cell, let's watch the
   clonal growth. Tumor terminated...good!
Starting with a single malignant cell, let's watch the
   clonal growth. Tumor terminated...good!
Starting with a single malignant cell, let's watch the
   clonal growth. 3 1 3 5 3 1 1 Tumor terminated...good!
Starting with a single malignant cell, let's watch the
   clonal growth. 4 6 5 6 3 3 1 Tumor terminated...good!
Starting with a single malignant cell, let's watch the
   clonal growth. 2 2 5 3 2 2 6 5 6 5 4 2 1 Tumor
   terminated...good!
Starting with a single malignant cell, let's watch the
   clonal growth. 3 5 3 7 6 3 3 1 1 Tumor terminated...good!
   I've seen enough!
```

cycle, the script counts the number of surviving cells. If the founding cell happens to die during the first cycle, the colony dies at the very first cycle. Through ten simulations we see that in most instances, a colony will die after about a half-dozen cell cycles. However, sometimes, the colony is lucky and continues to grow and grow. Simulations show that once a colony becomes large, its growth becomes exponential and seems to continue its growth forever (but not really). Probabilistic models such as this are called Monte Carlo simulations. For probability theorists, this is all familiar territory. For biologists who know a little Perl, these simulations provide an opportunity to model complex events that obey simple laws.

How is this actually accomplished in Perl? A short snippet of Perl contains the heart of the simulation (List 14.3.3).

LIST 14.3.3. PERL SNIPPET SHOWING THE ALGORITHM THAT REPEATEDLY ASSIGNS PROBABILISTIC OUTCOMES TO AN EVENT

```
while ($i < $sum +1)
   {
   $i++;
   $randnum = int( rand(100) ) + 1;
   if ($randnum > (100 * $value))
      {
      $sum = $sum + 1;
      }
   if ($randnum < ((100 * $value) -1))
      {
      $sum = $sum - 1;
      }
   }
```

In this snippet, Perl picks a random number from 1 to 100:

```
$randnum = int( rand(100) ) + 1;
```

Perl then compares that number with the preassigned (model) probability that an event will occur. In this case, the event is the death of a dividing cell. Let us say the preassigned probability that a cell will die during a growth cycle is 0.45. The variable *$value* is assigned 0.45 and multiplied by 100 to produce the number 45. If the number (from 1 to 100) that Perl randomly chose is greater than 45, then the model would let the cell live and increase the number of cells in the population by one (because the cell successfully divides if it survives).

```
if ($randnum > (100 * $value))
   {
   $sum = $sum + 1;
   }
```

If the number (from 1 to 100) that Perl randomly chose is less than 46, then the model would kill the cell and decrease the number of cells in the population by one (because the dead cell dropped from the population).

```
if ($randnum < ((100 * $value) -1))
   {
   $sum = $sum - 1;
   }
```

Perl repeats this exercise for every cell in the population and for every cell cycle for the duration of the simulation. The model permits us to change the cell death probability over and over again, to see how perturbations in cell death rate might alter population growth. Does the model accurately predict the growth of tumor populations from single cells? Not at all. The model is based on unproven assumptions about how cell populations grow. The purpose of the simulation is to let us see how changes in a variable might alter the outcome of complex events. Humans cannot, through any act of mentation, calculate the outcome of complex events. The model gives us some idea of the range of outcomes that might be expected under a set of assumptions. This knowledge can be used to build more sophisticated models based less on assumption and more on experimental evidence.

As an exercise for the reader, we include the script `monte6.pl`, which simulates the occurrence of metastases in a cell population. It determines whether metastasis has occurred when the tumor has reached a size of 50 million cells (List 14.3.4).

LIST 14.3.4. PERL SCRIPT `monte6.pl` MODELS METASTASES IN A GROWING CLONAL POPULATION

```
#!/usr/bin/perl
my $badprevchance =  0.00000009;
my $stillbad = $badprevchance;
my $goodprevchance = 0.000000009;
my $stillgood = $goodprevchance;
my%badhash;
my%goodhash;
my $size = 1;
my $i = 1;
my%count = 0;
my $badtotal = 0;
my $nicetotal = 0;
$badhash{$size} = $badprevchance;
while ($i < 30)
  {
  $size = $size * 2;
  my $baddoublechance = 1 -
  ((1-$badprevchance)*(1-$badprevchance));
  $badhash{$size} = $baddoublechance;
  $badprevchance = $baddoublechance;
  $i = $i + 1;
  }
$size = 1;
$i = 1;
```

```
$goodhash{$size} = $goodprevchance;
while ($i < 30)
  {
  $size = $size * 2;
my $gooddoublechance = 1 -
   ((1-$goodprevchance)*(1-$goodprevchance));
  $goodhash{$size} = $gooddoublechance;
  $goodprevchance = $gooddoublechance;
  $i = $i + 1;
  }
while ($count < 5000)
  {
  my $n = 1;
  $count++;
  my $bad = 1;
  my $nice = 1;
  my $badoutcome = "No mets for the bad tumor";
  my $niceoutcome = "No mets for the slow tumor";
  my $niceresult = 0;
  my $badresult = 0;
  my $randomvar = 0;
  while (1)   #this will loop forever
    #unless something in the block prompts an exit
    {
    $nice = 2 * $nice; #tumor number doubles
    if ($niceresult != 1)
    #if a metastatic state has not been achieved

       {
       $randomvar = rand();
       $niceresult = $randomvar < $goodhash{$nice} ? 1: 0;
       }
    if (($niceresult == 1) && ($nice < 50000000))
       {
       $niceoutcome = "$n $nice $niceresult";
       }
    $bad = 2 * $bad;
    if ($badresult != 1)
        {
        $randomvar = rand();
        $badresult = $randomvar < $badhash{$bad} ? 1: 0;
        }
    if (($badresult == 1) && ($bad < 50000000))
```

```
        {
        $badoutcome = "$n $bad $badresult";
        }
    $n = $n +1;
    last if (($bad > 50000000) && ($nice > 50000000));
    last if (($badresult ==1) && ($nice > 50000000));
    last if (($bad > 50000000) && ($niceresult == 1));
    }
  #print "$badoutcome\n$niceoutcome\n\n";
  $badtotal++ if ($badoutcome =~ /\d/);
  $nicetotal++ if ($niceoutcome =~ /\d/);
  }
print "$count $badtotal $nicetotal\n";
$chanceratio = ($stillbad) / ($stillgood);
print "chance ratio is $chanceratio\n";
$simratio = $badtotal / $nicetotal;
print "simulation ratio is $simratio\n";
exit;
```

14.4. ROUGH TEST OF THE BUILT-IN RANDOM NUMBER GENERATOR

Perl's built-in random number generator is not truly random. Eventually, the sequence of random numbers will repeat. When that happens, randomness is lost. For many purposes, such as simple Monte Carlo simulations, this will have no detrimental effect. For robust cryptographic work, a lame random number generator might be catastrophic.

Several tests for randomness have been proposed. Compression techniques all take advantage of the non-random nature of information (for example, text, images). Compression algorithms fail to compress strings composed of a truly random sequence of characters. If a file fails to compress (become smaller) with a compression utility, it is an excellent indicator that the file is composed of random characters.

A very crude indicator of randomness is an unbiased distribution of randomly chosen items. After a billion random selections from the alphabet, one might expect an approximately equal number of selected a's, b's, and c's, for example. A random selector producing a billion m's (and nothing else) would qualify as a very odd example of randomness.

The following snippet of code repeatedly selects integers from 1 to 100 and counts the number of times each integer is selected. All random number algorithms should yield a near-equal number of selections of each integer (List 14.4.1).

LIST 14.4.1. PERL SNIPPET TO DETERMINE UNBIASED RANDOM SELECTION

```
open (HOLD,">holder.txt")||die"cannot";
while ($n < 1000000)
  {
  $x = int(rand(100)) + 1;
#pick a number between 1 and a hundred
#make a hash of the numbers picked and the
#number of times each is picked
  $randhash{$x}++;
  $n++;
  }
foreach $key (sort byval keys %randhash)
  {
  print HOLD "$randhash{$key} $key\n";
  }
sub byval
  {
  $randhash{$a} <=> $randhash{$b};
  }
```

14.5. HOW OFTEN CAN I HAVE A BAD DAY?

Imagine this scenario. One of the hospital's pathologists has just made a diagnostic error on each of three consecutive reports. Luckily, the errors were detected by the pathology review conference and corrected before the reports were released.

The chair of the department calls the pathologist to her office and berates her for a completely unacceptable error rate. No pathologist should be permitted to diagnose three consecutive cases incorrectly.

The pathologist defends herself, saying that a long-term review of her cases shows that she has a 2 percent error rate, which is the national average for pathology errors. She cannot explain why three errors occurred consecutively, but she supposes that if you sign out enough cases, eventually you will make three consecutive errors. The chair is not persuaded by her explanation.

Who is right? A few lines of Perl can resolve the issue (List 14.5.1).

The Perl script simulates 100,000 diagnoses, which is a fair estimate of the total number of diagnoses a pathologist might render in their entire career (at 4,000 diagnoses per year over 25 years of service). Each diagnosis is assigned a random number between 0 and 100. The "diagnosis" loop is repeated 100,000 times. In each loop, if the randomly

LIST 14.5.1. PERL SCRIPT run.pl USES RESAMPLING TO SIMULATE RUNS OF ERRORS

```
#!/usr/local/bin/perl
$errorno = 0;
while ($count < 100001)
  {
  $count++;
  $x = rand(100);
  if ($x < 2)
     #simulates a 2% error rate
     {
     $errorno++;
     }
  else
     {
     $errorno = 0;
     }
   if ($errorno == 3)
     {
     print "Uh oh. 3 consecutive errors\n";
     $errorno = 0;
     }
  }
  exit;
```

assigned number is less than 2, the pathologist's error number is incremented by 1. If the next diagnosis is randomly assigned a number greater than 2, the error number is dropped back down to 0 (that is, the diagnosis is correct and the run of errors is broken). If an error occurs on three consecutive occasions, the event is printed to the computer monitor (List 14.5.2).

In this trial of 100,000 diagnoses, using a 2 percent error rate, the modeled pathologist had two runs of three consecutive errors. Because 100,000 diagnoses roughly accounts for the number of diagnoses rendered by a pathologist in her entire career, one can say that she can be permitted a three-error day twice in her career.

LIST 14.5.2. OUTPUT OF run.pl

```
c:\ftp>perl run.pl
Uh oh. 3 consecutive errors
Uh oh. 3 consecutive errors
```

14.6. THE MONTY HALL PROBLEM: SOLVING WHAT WE CANNOT GRASP

Here is the legendary Monty Hall problem, named after the host of a televised quiz show where contestants faced a similar problem:

> "The player faces three closed containers, one containing a prize and two empty. After the player chooses, she is told that one of the other two containers is empty. The player is now given the option of switching from her original choice to one of the other closed containers. What should she do? Answer: Switching will double the chances of winning."

Marilyn vos Savant, touted by some as the world's smartest person, correctly solved the Monty Hall problem in her newspaper column. As a result, she received thousands of responses, many from mathematicians, disputing her answer.

This is one of those rare problems that seems to defy common sense. Personally, whenever I tackled this problem using an analytic approach based on probability theory, I came up with the wrong answer.

In desperation, I decided to forget about logic and simply simulate Monty Hall's game, in Perl. It can be done with a two-person model. One person randomly chooses a place to hide the prize and the second person goes through the steps of picking the container. You can repeat this yourself a few thousand times, accumulating sufficient data to confirm or invalidate Marilyn vos Savant's solution.

Or, you can model the game with Perl, using Perl's built-in random number generator to hide the prize in one of three places (Lists 14.6.1, 14.6.2, and 14.6.3).

The script `montesw.pl` simulates the Monty Hall strategy where the player takes the option to switch her selection. The second script, `monteno.pl`, simulates the Monty Hall situation where the player declines to switch her selection. By taking the switch option, she wins nearly twice as many simulations as she would have won if she declined the switch. The beauty of the resampling approach is that the programmer does not need to understand why it works. The programmer only needs to know how to use a random number generator to create an accurate simulation of the Monty Hall problem that can be repeated thousands and thousands of times.

How does the Monty Hall problem relate to biomedicine? Preliminary outcomes of clinical trials are often so dramatic that the trialists choose to redesign their protocol midtrial. A drug or drug combination might have demonstrated sufficient effectiveness to justify moving control patients into the treated group. Or, adverse reactions might necessitate switching patients off a certain trial arm. In either case, the decision to switch protocols based on midtrial observations is a Monty Hall scenario.

LIST 14.6.1. PERL SCRIPT `montesw.pl` **SIMULATES THE SWITCH OPTION FOR THE MONTY HALL PROBLEM**

```
#!/usr/bin/perl
#Monty Hall simulation when you switch boxes
$count = 0;
$winner = 0;
while ($count < 10000)
  {
  @box_array = (1,2,3);
  $full_box = (int(rand(3))+1);
  #this determines the actual full box
   $guess_box = (int(rand(3))+1);
   #this is the box that you picked as first choice
   @empty_array = grep{$_ != $full_box} @box_array;
   #these are the two empty boxes
   #print "@empty_array\n";
   @showbox = grep{$_ != $guess_box} @empty_array;
   #these are the empty box(es)
   #with the guess box excluded
   if (scalar(@showbox) == 1)
   #if the guess box was the wrong choice
     {
     $winner++;  #we've decided that we will switch
     }
   $count++;
  }
print $winner;
exit;
```

LIST 14.6.2. PERL SCRIPT `monteno.pl` **SIMULATES THE NO-SWITCH OPTION FOR THE MONTY HALL PROBLEM**

```
#!/usr/bin/perl
#Monty Hall simulation when you refuse to switch boxes
$count = 0;
$winner = 0;
while ($count < 10000)
  {
```

```
  @box_array = (1,2,3);
  $full_box = (int(rand(3))+1);
  #this determines the actual full box
  $guess_box = (int(rand(3))+1);
  #this is the box that you picked as first choice
  @empty_array = grep{$_ != $full_box} @box_array;
  #these are the two empty boxes
  @showbox = grep{$_ != $guess_box} @empty_array;
  #these are the empty box(es)
  #with the guess box excluded
  if (scalar(@showbox) == 2)
  #if the guess box was the wrong choice
     {
     $winner++;  #we've decided that we won't switch
     }
 $count++;
  }
print $winner;
exit;
```

LIST 14.6.3. OUTPUT OF montesw.pl **AND** monteno.pl

```
C:\ftp>perl montesw.pl
6598
C:\ftp>perl monteno.pl
3408
```

CHAPTER 15

Bioinformatics

15.1. BACKGROUND

Definitions abound for the field of bioinformatics. Many workers in the profession have a focused definition, one involving the organization and analysis of large biomolecular datasets for the purpose of answering questions regarding basic biology. This definition encompasses the work actually performed by professional bioinformaticians and serves to distinguish the field of bioinformatics from the field of biomedical informatics (which uses computational methods to solve medical problems).

In the field of bioinformatics, Perl has held an impressive lead among programming languages. In the author's opinion, Java and Python are distant runners-up. The eminence of Perl in bioinformatics is usually credited to Perl's fast parsing routines that facilitate quick traversal and sequence matching through large DNA files. Probably more important is felicitous timing. The Human Genome Project was established by Congress, starting with the 1988 budget. The year 1988 was also when Larry Wall released the first version of Perl. The rising popularity of Perl closely matched the rising popularity of DNA sequence databases. In 2000, a group of Perl programmers and bioinformaticians released the first version of the open-source BioPerl project. BioPerl provides an interface to public genomic datafaces and implements a variety of supportive analytic algorithms. The entire BioPerl library can be installed directly through the ActiveState package manager or downloaded directly from the BioPerl site (*www.Bioperl.org*).

There is a wealth of high-quality, readily available information about Perl and genome sequence databases, and the reader is urged to consult other books if bioinformatics is their primary interest.

The purpose of this short chapter is to describe some freely available bioinformatics-related Perl libraries and modules and to provide a few sample Perl scripts for bioinformatics.

15.2. PERL MODULES FOR BIOINFORMATICS

LIST 15.2.1. SOME PERL MODULES FOR BIOINFORMATICS

```
AI-Gene-Sequence
AI-Genetic
Bio-Das
Bio-FASTASequence-File
Bio-FASTASequence
Bio-FastaStream
Bio-MAGE
Bio-MCPrimers
Bio-Phylo
Bio-ProteinFeatures
Bio-SAGE-Comparison
Bio-SAGE-DataProcessing
Bio-Tools-DNAGen
bioperl
DNA
FASTAParse
FASTASequence
GO-TermFinder
```

15.3. BIOPERL PROJECT

In 2000, the first version of the open-source BioPerl project was released. BioPerl provides an interface to public genomic datafaces and implements a variety of algorithms relevant to sequence analysis.

The BioPerl Project is an international association of developers of open-source Perl tools for bioinformatics, genomics, and life-science research.

If you have the ActiveState Perl interpreter, you can easily install the full BioPerl (62) library by invoking the Perl package manager:

```
c:\>ppm
ppm> install bioperl
```

The BioPerl modules are also available from CPAN as a compressed tar-archived file (63).

The BioPerl organization provides a detailed tutorial covering the latest software release:

http://bioperl.org/Core/Latest/bptutorial.html

The Pasteur Institute also hosts an excellent online BioPerl tutorial at:

http://www.pasteur.fr/recherche/unites/sis/formation/bioperl/

Documentation and code for all modules and methods are available at:

http://doc.bioperl.org/releases/bioperl-1.0.1/

List 15.3.1 is a simple script that gets a sequence from a public database.

LIST 15.3.1. PERL SCRIPT `seq.pl` FOR FETCHING DATA FROM A GENOMIC DATABASE

```
#!/usr/local/bin/perl
use Bio::Perl;
use Data::Dumper;
open (TEXT,">seq.out");
$seq_object = get_sequence('swiss',"ROA1_HUMAN");
print TEXT Dumper($seq_object);
write_sequence(">roa1.fasta",'fasta',$seq_object);
write_sequence(">roa1.raw",'raw',$seq_object);
exit;
```

15.4. FINDING PALINDROMES IN A GENE DATABASE

Readers may want to try their hand writing scripts that parse through sequence data. Palindromes are an easily understood entry point to this subject. Palindromes are character strings that read the same forward and backward (for instance, bob, otto, civic). A DNA palindrome is a sequence for which the header bases are the reverse complement of the tail bases. For instance GAATTC is a palindrome. To understand why this is so, you must know that DNA is composed of sequences of four bases, Guanine (G), Adenine (A), Thymine (T), and Cytosine (C), and that each strand of DNA is matched by a complementary second strand in which G on the either strand is always coupled with C on the other strand, and A on the either strand is always coupled with T on the other strand.

A simple DNA palindrome is GAATTC:

```
G at front complements C at end
A second from front complements T second from end
A third from front complements T third from end
```

Notice that the sequence GAATTC is the same as its complementary strand read backwards. Other examples of short DNA palindromes are AAG_CTT, and GCCC_GGGC. Another type of DNA palindrome allows for spacer regions with one or more non-palindromic bases separating the reverse-complement sequence. Spacer-separated reverse-complement sequences would permit looped strand self-annealing. By any definition, DNA palindromes are unusual sequences characteristic of a given gene.

Relevant literature describes the role of palindromic sequences as transcription-binding sites and their association with gene amplification and genetic instability.

A Perl script that can find gene palindromes, including reverse-complement non-repeating palindromes with an intervening spacer region is shown in List 15.4.1. The example script operates on a given uninterrupted sequence of C, A, G, T combinations, held in the sequence file "sample" (List 15.4.2). This algorithm can be modified to find any type of sequence palindrome, indexing all found algorithms, and parsing through gene sequence data of any length. A modified script found all simple palindromes in a 34+ megabyte sequence file, in just 411 seconds on a 1.6 Ghz CPU (List 15.4.3).

LIST 15.4.1. PERL SCRIPT `palindro.pl` FINDS PALINDROMES IN A GENE SEQUENCE

```
#!/usr/bin/perl
$filename = "sample";
#load a file consisting of a DNA sequence
open (TEXT, "sample")||die"Cannot";
$line = " ";
$count = 0;
for $n (5..20)
    {
    $re = qr /[CAGT]{$n}/;
    $regexes[$n-5]= $re;
    #creates an array of regex patterns matching any
    #DNA sequence from 5 bases up to 20 bases.
    #Utilizes the qr() operation that compiles the pattern
    #once so that it needn't be re-compiled with each
    #iterative call to the pattern
    }
NEXTLINE: while ($count < 1000)
    {
    $line = <TEXT> ;
```

```
    $count++;
    foreach my $value (@regexes)
        #repeats for all the sequence patterns of
        #length 5 up to 20
        {
        $start = 0;
        while ($line =~ /$value/g)
            #repeats for every occurrence of the pattern length
            #in the sequence
            {
            $endline = $';
            $match = $&;
            $revmatch = reverse($match);
            #reverses the sequence of $match
            $revmatch =~ tr/CAGT/GTCA/;
            #and then substitutes the base complements
            #for each base in the reversed sequence.
            #This yields a palindromic tail for a
            #matched sequence
            if ($endline =~ /^([CAGT]{0,15})($revmatch)/)
                #for any original match, allow any intervening
                #sequence of length 0 to 15 followed by the
                #palindromic matching sequence
                {
                $start = 1;
                $palindrome = $match . "*" . $1 . "*" . $2;
                $palhash{$palindrome}++;
                }
            }
        if ($start == 0)
            {
            goto NEXTLINE;
            }
        }
    }
close TEXT;
while(($key, $value) = each (%palhash))
  #print out all the palindromes held in the associative array
    {
    print "$key => $value\n";
    }
exit;
```

LIST 15.4.2. EXCERPTED INPUT OF `palindro.pl` (LINE BREAKS OMITTED FROM ORIGINAL FILE)

```
ATGAGCGAAGAAAGCTTATTCGAGTCTTCTCCACAGAAGATGGAGTACGAAATTACAAAC
TACTCAGAAAGACATACAGAACTTCCAGGTCATTTCATTGGCCTCAATACAGTAGATAAA
.
.
.
.
AAGATCAGAAGCGACCATGACAATGCTATTGATGGATTATCTGAAGTTATCAAGATGTTA
TCTACCGATGATAAAGAAAAATTGTTGAAGACTTTGAAATAA
```

LIST 15.4.3. EXCERPTED OUTPUT OF `sample.pl`

```
(* separates the spacer region from the flanking palindromic
   regions)
C:\FTP>perl sample.pl
CTTTG*TCAGGATGGGC*CAAAG => 1
AGTAT*T*ATACT => 1
GAAATC**GATTTC => 1
AGTTT*GGCATCC*AAACT => 1
CCTTA*CCCTGT*TAAGG => 1
CTTCT*GGAGATTGAGA*AGAAG => 1
.
.
.
GATGG*ATTCAAG*CCATC => 1
GTTTGG*CAT*CCAAAC => 1
CTTCT*CCAC*AGAAG => 1
```

Palindromes can be quickly retrieved from large sequence datasets and used to annotate genes. Palindrome annotation is one of many ways to link raw sequence data to gene signature data, biological feature data, and (ultimately) pathologic lesions.

15.5. CLUSTERING: ALGORITHMS THAT GROUP SIMILAR OBJECTS

A common task in biomedicine is to group items based on their similarity. This might involve grouping histopathologic images based on similar content, organizing documents based on content, or clustering samples of diseased tissues based on shared genetic or proteomic profiles. This is particularly useful when a sample is characterized by a multiplex profile consisting of hundreds or thousands of individual measurements (for example, gene expression levels, protein levels, wave amplitudes, physical attributes).

Michiel de Hoon and his co-workers have prepared a library of programs that will cluster sets of data based on a variety of similarity algorithms (64). They have prepared a Perl implementation that can be easily installed with the ActiveState Perl package manager:

```
c:\>ppm install http://bonsai.ims.u-
tokyo.ac.jp/~mdehoon/software/cluster/Algorithm-Cluster.ppd
```

An example using their Perl module `Algorithm::Cluster` is shown (List 15.5.1) The output also is shown (List 15.5.2).

LIST 15.5.1. PERL SCRIPT `cluster.pl`, A PERL SCRIPT DEMONSTRATING A CLUSTERING ALGORITHM

```
#!/usr/local/bin/perl
use Algorithm::Cluster;

$row0 = [ 3, 4, 1, 3, 4, 7, 5];
$row1 = [ 1, 2, 32, 15, 60, 76, 87];
$row2 = [ 1, 2, 32, 13, 60, 80, 87];
$row3 = [ 5, 6, 3, 2, 5, 9, 2];
$row4 = [ 5, 6, 3, 99, 5, 9, 2];
$row5 = [ 5, 6, 3, 115, 5, 9, 2];
$data = [ $row0, $row1, $row2, $row3, $row4, $row5 ];
my %param = (
nclusters => 3,
data => $data,
mask => '',
weight => '',
transpose => 0,
npass => 10,
method => 'a',
dist => 'e',
initialid => [],
);
my ($clusters, $error, $found) =
   Algorithm::Cluster::kcluster(%param);
$i = 0;
foreach my $thing (@$clusters)
   {
   print "Row" . "$i" . " => Cluster $thing\n";
   $i++;
   }
exit;
```

LIST 15.5.2. OUTPUT OF `cluster.pl` SCRIPT

```
c:\ftp>perl cluster.pl
Row0 => Cluster 1
Row1 => Cluster 2
Row2 => Cluster 2
Row3 => Cluster 1
Row4 => Cluster 0
Row5 => Cluster 0
```

The script selected three sets of similar clusters from the test set of six data arrays:

Cluster 0:

$row4 = [5, 6, 3, 99, 5, 9, 2];

$row5 = [5, 6, 3, 115, 5, 9, 2];

Cluster 1:

$row0 = [3, 4, 1, 3, 4, 7, 5];

$row3 = [5, 6, 3, 2, 5, 9, 2];

Cluster 2:

$row1 = [1, 2, 32, 15, 60, 76, 87];

$row2 = [1, 2, 32, 13, 60, 80, 87];

The script clustered six arrays of data into three groups based on the similarity of their contained data elements. This script was written to conform to the usage syntax included in the distributed Perl module, `Algorithm::Cluster`. Readers will notice some unfamiliar syntax in the script. Straight brackets (instead of parentheses) enclose arrays of data and an array variable has a strange prefix, (`@$cluster`). The module makes liberal use of Perl references, which are variables that contain the location in which data resides. References have their own syntax by which they are created and from which their referred data can be retrieved. We will not explain Perl reference syntax for two reasons: 1) the syntax for reference calling will change in the next version of Perl; and 2) object-oriented programming (see Glossary) techniques can eliminate the need to directly de-reference encapsulated object data. A casual inspection of the row data should convince the reader that the script can cluster arrays of numbers based on their similarities.

CHAPTER

16

Network Computing

16.1. BACKGROUND

Thus far, the emphasis of this book has been on self-reliance. Many informatics tasks can be done on a personal computer, if the user has access to relevant datasets. Access to datasets usually involves the most basic Internet resources: an e-mail account and a Web browser. Sometimes the most effective way to advance an informatics agenda is through networking. A succession of increasingly complex networking protocols characterizes the evolution of network-based computational tasking (List 16.1.1).

LIST 16.1.1. INCREASINGLY COMPLEX TASK-SHARING NETWORK PROTOCOLS AND METHODS

1. FTP (file transfer protocol)
2. TELNET
3. HTTP (hypertext transfer protocol)
4. CGI (Common Gateway Interface)
5. RPC (remote procedure calls)
6. XML-RPC (XML-based remote procedure calls)
7. SOAP (simple object access protocol)
8. P2P (peer-to-peer networking)
9. WEB Services
10. GRID computing

A full discussion of these protocols is beyond the scope of this book. Readers should know that the popular scripting languages provide open-source modules for the less complex protocols (FTP, TELNET, HTTP, RPC, and SOAP).

The purpose of this chapter is to acquaint the reader with a few Perl options for network computing.

16.2. PERL INTERNET MODULES

Several books are dedicated to describing available Perl modules (65), (40) (List 16.2.1). Motivated readers are encouraged to implement these useful protocols.

LIST 16.2.1. PERL MODULES FOR WEB PROGRAMMING AND DISTRIBUTED COMPUTING

```
Acme-spider
CGI
CGI-Widget
CGI-Wiki
HTTP-file
HTTP-getimages
HTTP-headers-useragent
HTTP-lite
HTTP-messageparser
HTTP-mobileagent
HTTP-parser
HTTP-server-simple
HTTP-simplelinkchecker
IO-ftp
IO::socket
Lib::www
LWP-protocol-sftp
Net-ftp-file
Net-ftp-recursive
Net-telnet
RPC-xml
SOAP-message
SOAP-mime
SOAP-wsdl
WWW-search
```

16.3. FTP (FILE TRANSFER PROTOCOL)

FTP is an Internet protocol for transferring files between different networked computers. Files can be uploaded or downloaded. Most people's experience with FTP is limited to FTP client applications that access an FTP site, browse through the FTP subdirectories, and then download selected files (from the FTP server to their hard drive).

Perl has a bundled module that supports the four most important FTP functions: login (get into the FTP site), cwd (change directories), get (download [List 16.3.1] a file), and put (upload a file in upload-tolerant sites).

LIST 16.3.1. PERL SCRIPT `ftp_omim.pl`, TO DOWNLOAD THE OMIM FILE

```
#!/usr/local/bin/perl
use Net::FTP;
#Net::FTP comes bundled in Perl distributions
$host = "ftp\.ncbi\.nih\.gov";
#We will download a public file from the NCBI site
$user = "anonymous";
$password = "guest";
#This site expects a username of anonymous and
#a password of guest
$dir = "\/repository\/OMIM";
#We will go to the repository/OMIM subdirectory
$file = "omim\.txt\.Z";
#What we want is a 49 Megabyte gzip-compressed file named omim.txt.Z
$ftp = Net::FTP->new($host);
#We make a new FTP object
#Once we have the object, we can call the module methods
$ftp->login($user, $password);
$ftp->cwd($dir);
#We will go to the desired subdirectory
$ftp->binary;
#This is a compressed file, and it should be
#downloaded as a binary file. If it were a plain-text
#or html file, we might have downloaded it with the ascii method
$ftp->get($file, $file);
#The get command downloads the file and gives the file
#the same name, on our hard drive
$ftp->quit;
#all done
end;
```

This basic script can easily be modified to download/upload whole subdirectories from multiple FTP sites. In this instance, we downloaded a single file, `omim.txt.Z`, the compressed OMIM (Online Mendelian Inheritance in Man, see Appendix). Save this file. We will be using it in Chapter 19.

16.4. SOME NETWORK COMPUTING DEFINITIONS

Telnet is a network protocol for establishing command-line sessions on remote host computers.

HTTP is the familiar World Wide Web protocol that permits Web pages to be exchanged. Enhancements of server and client functionality (through CGI scripts on the server side and applets on the client side) empower users to distribute computational tasks using the HTTP protocol.

RPC is remote procedure calling. A client invokes a computational method and calls another networked computer to execute the method.

XML-RPC is an RPC call performed using XML-configured commands.

SOAP is a formalized protocol that establishes uniform terms and methods for describing RPC transactions in XML syntax.

Web Services are resident server functionalities that accept client-initiated requests through an interface described by WSDL (Web Services Descriptive Language).

P2P is a technique for distributing messages (queries and responses) through a network of peers. Peers are networked host computers that, for purposes of the P2P protocol, are somewhat autonomous and have properties that are neither client-type nor server-type.

GRID computing is a method for providing Web services by brokering client requests through a network of participating servers. In the most advanced grid computing architecture, requests can be broken into computational tasks that are processed in parallel on multiple computers and transparently (from the client's perspective) assembled and returned.

Web Services, P2P, and GRID require complex architectures. Projects based on these technologies are typically developed by groups of computer scientists blessed with time and expertise.

16.5. MAILINGS

Occasionally, you might want to do an e-mailing from a list of e-mail addresses or a combined list of e-mail addresses and corresponding addressee names. In Chapter 3 Section 14 we learned how easily you can extract an unlimited number of e-mail addresses and names via a PubMed search. The following script is a standard Perl mailing script that parses a list of e-mail addresses and sends each address a separate e-mail message (List 16.5.1).

LIST 16.5.1. PERL SCRIPT `sendthem.pl` **SENDS AN E-MAIL MESSAGE TO EACH MEMBER OF A LIST**

```
#!usr/bin/perl
$recipient = " ";
open(NAME, "names.txt");
#names.txt is a file-list of e-mail addresses
while ($recipient ne "")
     {
     $recipient = <NAME>;
     $recipient =~ s/\n//o;
     $mail{From} = YOUR EMAIL ADDRESS HERE;
     $mail{To}   = $recipient;
     $server = YOUR SERVER HERE;
     use Mail::Sendmail;
     $loaded = 1;
     if ($server)
         {
         $mail{Smtp} = $server;
         print "Server set to: $server\n";
         $mail{Subject} = "Test";
         $mail{Message} = "Hello\n\n";
         print "Sending...\n";
         }
     if (sendmail%mail)
        {
        print "\$Mail::Sendmail::log:\n$Mail::Sendmail::log\n";
        if ($Mail::Sendmail::error)
           {
           print "\$Mail::Sendmail::error:\n$Mail::Sendmail::error\n";
           }
        print "ok 2\n";
        }
     else
        {
        print "\n!Error sending mail:\n$Mail::Sendmail::error\n";
        print "not ok 2\n";
        }
   }
exit;
```

In the sendthem.pl script, a file containing a list of e-mail addresses is parsed, and each loop of the script sends an e-mail to a different mail recipient. This script is particularly useful when you have a large mailing list or when you prefer that each recipient receive her own message (instead of listing every recipient on the same "To:" line). Be sure to include your own e-mail address as the sender (the $mail{From} line). Include your mail server name on the $server *line.* Most e-mail applications have an option named "Account Settings" or "Server Settings" in which the mail server name is listed.

As the sendthem.pl script runs, it produces a running commentary on the sender's monitor. For example:

```
c:\ftp>perl sendthem.pl
Server set to: smtp.someplace.net
Sending...
$Mail::Sendmail::log:
Mail::Sendmail v. 0.79 - Tue Nov 21 12:58:25 2006
Date: Tue, 21 Nov 2006 12:58:25 -0500
Server: smtp.someplace.net Port: 25
From: jjberman1@someplace.net
Subject: Test
To: jjberman1@someplace.net
Result: 250 ok ; id=20061121175824m1200ktvo5e
```

Readers are warned that this script is powerful. Responsible users will test the script before sending a list of e-mails, send messages to a small number of recipients, use their own address as the sender, and closely monitor the output from each sendmail mailing.

16.6. CLIENT-SERVER ON A SINGLE COMPUTER WITH PERL

When you sit at home or in the office and call Web pages in your browser, your computer is acting as the client. The URL that sends you your Web page is acting as the server in a client-server transaction.

Most people do not know that if you are connected to the Internet, your personal computer can act as a server or a client, or both at once.

In Chapter 9, Sections 5 and 6, we discussed Perl CGI scripts. The examples all required access to a server-side cgi-bin subdirectory. In this chapter, we will show that you can write your own client and server scripts, and test them both simultaneously on your own computer!

The demonstration program will consist of a Web page that includes a form that accepts a user query. The Web page will send the query to a designated server (in this case, the user's own computer). The server will extract the query and collect all of the lines from a text file that contains the query word. The resultant list of lines will be sent through the Internet to the client's browser as a Web page reply. All of this will be done with a few dozen lines of Perl code.

Let us begin with the code for a short Web page that provides a box into which the user may enter a query word (List 16.6.1).

LIST 16.6.1. SIMPLE WEB PAGE, `client.htm`, WITH QUERY FORM

```
<html>
<head>
<title>post</title>
</head>
<body>
<br>
<form name="sender" method="GET" action="http://127.0.0.1:1080">

<br><center>
<input type="text" name="tx" size=38 maxlength=48 value="">

<input type="submit" name="bx" value="SUBMIT">
<br><br>
</center></form>
<br>
</body></html>
```

The first two lines are standard HTML header and the last three lines are standard footer. The remaining three lines create the input form:

```
<form name="sender" method="GET"  action="http://127.0.0.1:1080">

 <br><center>
<input type="text" name="tx" size=38 maxlength=48 value="">

 <input type="submit" name="bx" value="SUBMIT">
```

Notice that the query will be submitted as a `GET` message, and the action will involve sending the `GET` message to server `"http://127.0.0.1:1080"`. The address 127.0.0.1 is a reserved address for the host computer. In Internet terms, it is equivalent to "myself." The suffix to the host address is "1080." This is a number that we have somewhat arbitrarily chosen as the server port. The created Web page is shown in Figure 16-1.

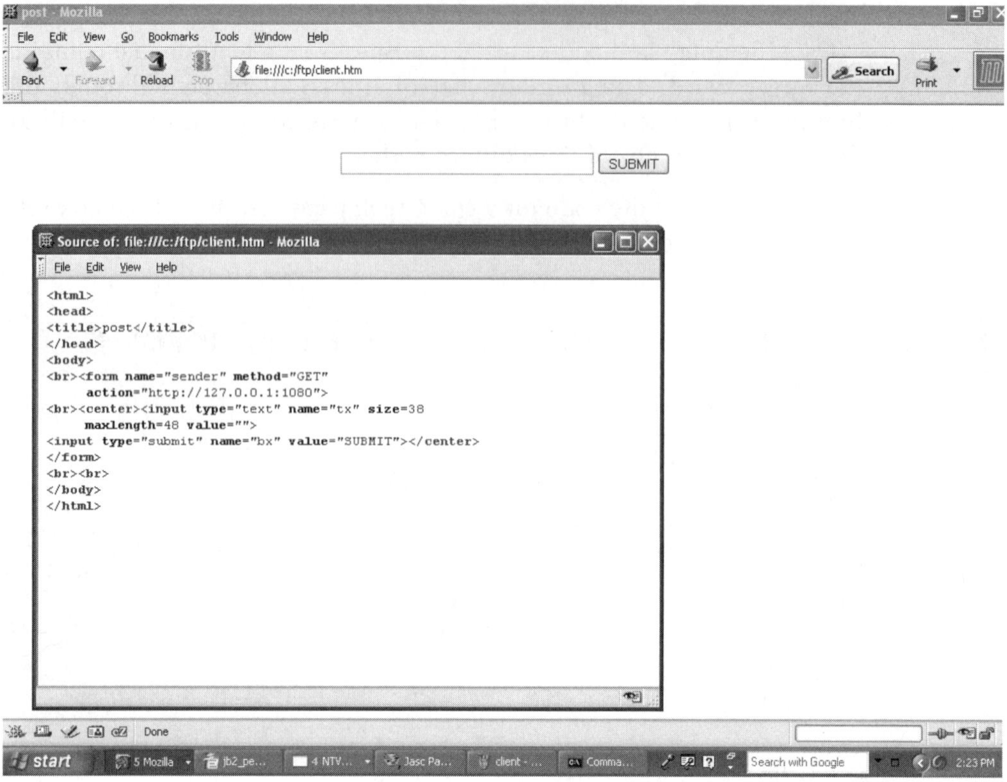

Figure 16-1 The Web client page, `client.htm`, containing a simple query input form.

Before we use the Web page to create and enter a query word and then submit the query, we need to have a server script that listens for our message.

The `server_4.pl` script uses the `IO::Socket` module, contained in standard Perl distributions, to receive the message and return a reply (List 16.6.2).

LIST 16.6.2. PERL SCRIPT `server_4.pl` REPLIES TO THE `client.htm` WEB PAGE

```
#!/usr/bin/local/perl
use IO::Socket;
$port = 1080;
my $sock = IO::Socket::INET->new(LocalAddr => '127.0.0.1',
                                 LocalPort => $port,
                                 Listen => 5,
                                 Proto => 'tcp',
                                 Reuse => 1,
                                );
die "Can't create listen socket: $!" unless $sock;
```

```
print "Listening for connections on $port\n";


while($new_sock = $sock->accept)
  {
  handle_connection($new_sock);
  $new_sock -> close;
  }


sub handle_connection
    {
    my ($s) = @_;
    print $s qq|\n <html><head><title>Post</title></head> |;
    print $s qq|\n <body><h2><center>Post</center></h2> |;
    if (defined ($buf = <$new_sock>))
       {
       #print $s "<br>$buf\n";
       if ($buf =~ /tx\=(.+)\&/)
          {
          $lookup = $1;
          print $s qq|<br><br><center><h3>$lookup</h3>
                      </center><br><br>\n\n\n\n |;
          }
       }
    open(TEXT,"how.out")||die"Cannot";
    my $line = " ";
    while ($line ne "")
        {
        $line = <TEXT>;
        if ($line =~ /$lookup/)
          {
          print $s "\<br\>$line\n";
          }
        }
    close TEXT;
    print $s qq|\n <br><br></body></html> |;
    print STDERR "Received connection from @{[$s->peerhost]}\n";
    print STDERR "Received connection from @{[$s->peerport]}\n";
    $s->close;
    print STDERR "Connection closed\n";
}
exit;
```

This short script uses standard code recommended in the `IO::Socket` module. The two lines that contain information specific for this script are:

```
$port = 1080;
my $sock = IO::Socket::INET->new(LocalAddr => '127.0.0.1',
```

Here, we set the port to 1080 (the same port designated in the client.htm Web page) and the server address as 127.0.0.1. This ensures that the client's Web page will send its `GET` message to our server.

The actual `GET` message sent by the server is contained in:

```
http://127.0.0.1:1080/?tx=string&bx=SUBMIT
```

The `GET` message is passed to the handle_connection subroutine through the *`$new_sock`* file handle.

The query string is the first matching parenthetical expression in the `regex` line:

```
if ($buf =~ /tx\=(.+)\&/)
```

Once the query string is extracted, the file `how.txt` (which happens to be the raw text file containing the contents of this book) is parsed. All lines matching the query word are sent to the reply Web page (see Figure 16-2).

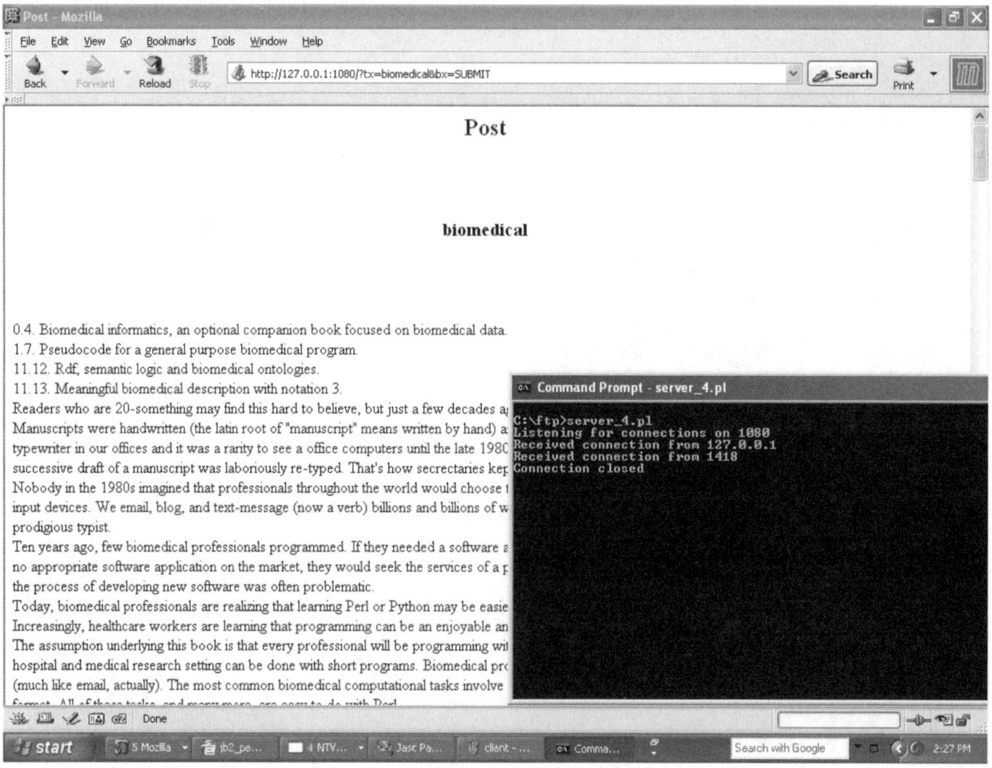

Figure 16-2 The Web page created by the server and sent to the client.

```
print $s "\<br\>$line\n";
```

Using Perl, it is easy to write and test powerful server-side scripts for handling input received via the Internet.

16.7. A WEB SERVICE RESOURCE FOR PERL

Web services are server-based collections of data, along with software methods that operate on the data. The data and software methods on the server can be accessed by remote clients. One of the features of Web services is that they permit client users (humans or software agents) to discover the kinds of data and methods offered by the Web service and the rules for submitting server requests.

With few exceptions, teams of software developers who have special expertise in WSDL (Web services description language, see Glossary) or GRID computing (see Glossary) prepare Web services. The novice programmer can be content in accessing Web services from Perl scripts.

A Web service of interest to biomedical professionals is the U.S. National Cancer Institute's caBIO:

http://ncicb.nci.nih.gov/core/caBIO/caBIO_FAQ

The caBIO services provide users access to cancer-related public databases and to a set of related methods. The Web site contains a description of methods and datasets.

Perl scripts for caBIO are also available from their Web site:

http://nc icb.nci.nih.gov/core/caBIO/technical_resources/code_examples

These public Perl scripts require that you have SOAP::Lite modules installed. At least one of the scripts requires that you have the DBI module installed. I have personally tried several of the public caBIO Perl scripts, and they operated as promised.

CHAPTER 17

A Quick Peek at Object-Oriented Programming

17.1. BACKGROUND

Object-oriented programming enjoyed a great deal of hype during the late 1990s and early 2000s. Those who understand the principles of object-oriented programming often become proselytizers of the methodology. The downside of object-oriented programming is that there is a very steep learning curve, and it might take years of devoted study and practice to master the technique. Most biomedical professionals simply do not have the time and energy to become object-oriented programmers.

The purpose of this chapter is to provide an overview of object-oriented programming and an explanation of how object-orientation is implemented in Perl. Readers should acquire sufficient knowledge to use freely available class libraries that are relevant to their professional interests.

17.2. OVERVIEW OF OBJECT-ORIENTED PROGRAMMING

There are a few tips that will help us safely deal with this challenging subject (List 17.2.1).

In object-oriented programming, methods (roughly equivalent to commands or to subroutines) belong to classes. The methods in a class are called by first declaring a variable to be a new object in the class. Once that's done, there is a syntax that allows the new class object to call upon any of the methods of its class or of the ancestor classes.

Used correctly, object orientation creates shorter, simpler, programs, reusable routines, and sensible software documentation. Used incorrectly, object orientation produces complex, unpredictable, and slow programs that cannot be de-bugged when awful and unexpected events occur (List 17.2.2).

LIST 17.2.1. TIPS ON DEALING WITH OBJECT-ORIENTED PERL

- Object-oriented Perl is a programming technique, not a new language. You don't need to install an "object Perl" interpreter or install new object-oriented modules. Object-orientation is always available for you either to use or to ignore in every Perl script you write.
- You can use the object-oriented Perl scripts (class modules) that other people have written, without knowing much about object-oriented Perl. This chapter will demonstrate how easily this can be done.
- Writing class modules is also easy, but writing well-designed class modules that successfully implement the best object-orientation features is very difficult. If your goal is to write short scripts to help you with your professional activities, object-oriented programming is neither necessary nor advisable.

LIST 17.2.2. PROBLEMS WITH OBJECT-ORIENTED PROGRAMMING

- Speed. Every time you call a class or external package, it slows your script.
- Readability versus comprehension. Object oriented scripts are usually short and easy to read. Unfortunately, you never quite know what the script is doing because the source code for the called methods reside in class libraries external to the script.

17.3. OBJECT-ORIENTED PROGRAMMING IN PERL

In a pure object-oriented language, methods are called by objects (usually, an instance of a class), all methods belong to some class, and everything (including every method) is an object. Perl is not a pure object-oriented language. In Perl, object orientation means that the programmer can create classes (usually stored in external scripts called modules) and that Perl scripts can declare new objects of these external classes. Once created, these objects have access to the methods contained in their specific classes as well as in the ancestor classes. In Perl, some data structures are not objects and some methods are not members of any class.

In most object-oriented programming languages, hundreds of class methods are bundled into the language. All of these methods can be immediately used in scripts. One of the difficult tasks in mastering an object-oriented language is learning what all the class methods do. This is sometimes referred to as learning the language's class library. Programmers can also create their own classes, class methods, instances of classes, and instance-specific methods. Programmers can also create subclasses of existing classes.

To use pre-existing class modules in a Perl script, you need to do three simple things (List 17.3.1).

LIST 17.3.1. THREE THINGS YOU MUST DO WHEN USING A PRE-EXISTING CLASS MODULE IN YOUR PERL SCRIPT

1. Start by telling Perl that you will be using a class module. and provide the directory location of the module if it is not installed in your Perl subdirectory.
2. Declare a new instance of the class. Have the class instance invoke the methods of the class.
3. Assign the results of the method to a variable.

The following short script implements the Sentence module, previously described in Chapter 3 Section 13 and Chapter 10 Section 4. We will review, in some detail, an object-oriented implementation of the module (List 17.3.2).

LIST 17.3.2. PERL SCRIPT `ob_trial.pl`, WHICH USES OBJECT-ORIENTED PROGRAMMING TECHNIQUE

```
#!/usr/local/bin/perl
use lib "c:/ftp/parse";
use Sentence;
$object1 = Sentence->new();
$filename = "testfile.txt";
$outputfilename = $object1->sentence_parse($filename);
print $object1->get_day();
exit;
```

The `ob_trial.pl` script demonstrates how to use the author's `Sentence.PM` module. Let us examine the script in some detail.

The first line begins like any other Perl script:

```
#!/usr/local/bin/perl
```

The second line invokes the lib module. The lib module tells Perl that when it looks for any module, it should include the specified subdirectory (c:\ftp\parse in this case) in its search path. Perl will re-orient the slashes for the appropriate operating system:

```
use lib "c:/ftp/parse";
```

The next line tells Perl that the script will use the Sentence module. As it happens, the Sentence module is located in the c:\ftp\parse subdirectory on my computer. So Perl would have no trouble finding the module:

```
use Sentence;
```

Almost every Class has a new() method because the primary purpose of any class is to create new instances of itself. We chose a name for our new object and created it using a special syntax:

```
$object1 = Sentence->new();
```

$object1 is a Perl variable unlike others. It is not a string, an array, or a hash. It is a Class object. In this case, it is an instance object of the class Sentence. It is entitled to call all of the methods written into the Sentence class and all of the methods inherited by the Sentence class (from its class ancestors).

The only difficult part of implementing existing classes is learning the methods included in the class. This can usually be achieved by reading the documentation that comes with the class module (see Figure 17-1).

The Sentence class documentation lists a class instance method named sentence_parse, which accepts the name of a file and parses the file, producing an output file that contains the sentences from the original file, assigning one sentence to each line in the output file. Such a method makes it easy for natural language parsers to evaluate the individual sentences in a text file. We call the Class method with the class object, as shown:

```
$outputfilename = $object1->sentence_parse($filename);
```

The method returns a file and a filename (*$outputfilename).*

Just for fun, we call the get_day() method, which is an instance method belonging to an ancestor class of Sentence:

```
print $object1->get_day();
```

The get_day method prints the date to the monitor. The sentence_parse and the get_day methods both send some output to the monitor (List 17.3.3).

Writing good class modules is much more difficult than using good class modules. Class module design is one of the most fascinating and chal-

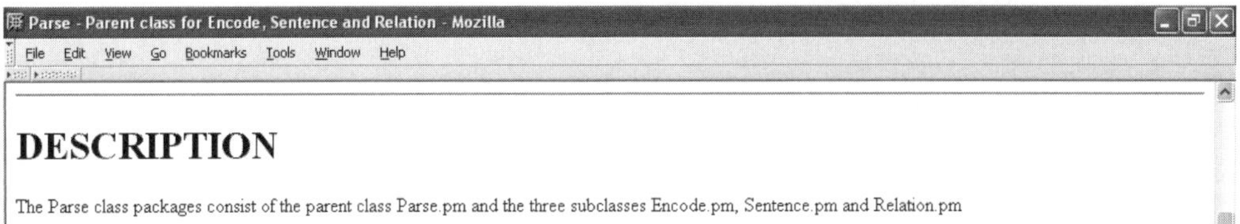

Figure 17-1 Documentation for the Parse class module, including the Sentence subclass.

lenging areas of Perl programming and will not be included in this book. Interested readers should consult Damian Conway's excellent book, *Object-Oriented Perl* (66).

Object-oriented programs can be very short if they draw primarily from class modules. The code that would normally be included within the script is transparently implemented by well-designed class methods.

LIST 17.3.3. OUTPUT OF FILE `ob_trial.pl`

```
C:\ftp\parse>perl ob_trial.pl
The file to be parsed is: testfile.txt
The name of your ouput file is RKCECWOC.NEI
Elapsed time = 0 seconds
This output was produced using Sentence perl class package on:
Sunday - June 4, 2006
```

CHAPTER

18

Use-Case: A Digital Image Specification in RDF

18.1. BACKGROUND

In Chapter 11, we introduced RDF (Resource Description Framework). RDF underpins the semantic Web, ontologies, semi-autonomous software agents, inference engines, and artificial intelligence. RDF is so future oriented that it is difficult to gain a realistic understanding of what to expect from RDF.

Aside from the jargon, RDF provides a simple method for specifying information as data triples. In Section 11.3, we contrasted standards with specifications. Much of the time and expense associated with developing and deploying data standards can be eliminated by a consistent implementation of recommended RDF data specification practices (67), (68). It is likely that in the near future, efforts to create new standards for biomedical data objects (for example, images, multiplex diagnostics tests, treatment guidelines, electronic laboratory notebooks, clinical trial datasets, instrument settings) will be supplanted by RDF specifications.

In my opinion, there simply is no technique in the field of biomedical informatics of greater importance than RDF. RDF is the key to all data organization, data sharing, and data standardization in the field of biomedical informatics. The RDF concepts first discussed in Chapter 11 are best understood by reviewing an example case. I was co-chair of the Laboratory Digital Imaging Project sponsored by the Association for Pathology Informatics, which developed an RDF schema for medical image objects. Many of the data elements and RDF schema examples were excerpted from work performed under this open access effort. Further information can be found at:

http://www.pathologyinformatics.org/

The purpose of this chapter is to demonstrate how Perl can implement an RDF data specification for any biomedical data object.

18.2. AN APPROACH TO DATA SPECIFICATION

An RDF data specification is basically an RDF schema that contains the classes and properties relevant to the domain of a data object (List 18.2.1).

LIST 18.2.1. EXAMPLES OF DATA OBJECTS IN BIOLOGY AND MEDICINE

- Flow cytometry report
- Chest x-ray
- Photomicrograph
- Operation note
- Karyotype report
- Quality assurance report
- Blood test
- Clinical trial
- Gene expression study

The RDF schema provides a formal vocabulary and a general syntax for describing included instances of data objects.. People who use an RDF data specification can merge their data with any other RDF document.

Creating an RDF data specification effort typically consists of just a few tasks (List 18.2.2).

LIST 18.2.2. TASKS TO CREATE AN RDF DATA SPECIFICATION

1. Prepare an RDF schema for the data object.
2. Recommend a method for assigning unique instances of the data object. In the case of a medical image, this would involve providing a unique name to the image so that it won't be confused with any other image.
3. Provide real-world examples of specified instances of the data object that people in the field can adopt in their own work.

18.3. CREATING AN RDF SCHEMA FOR A DATA OBJECT

An RDF schema contains the vocabulary that will be used in RDF documents to describe data objects in a particular domain. In this chapter we will be describing an RDF schema for the domain of medical images. Any person or user community can prepare an RDF schema that can be accessed by the world via the Internet (List 18.3.1).

LIST 18.3.1. GENERAL INSTRUCTIONS FOR CREATING AN RDF SCHEMA FOR A KNOWLEDGE DOMAIN OR FOR A CLASS OF DATA OBJECTS

1. List the classes and properties that comprehensively describe a knowledge domain.
2. Determine a class hierarchy. Every class on your list must be a subclass of something else. The top-level classes may be subclasses of Class.
3. Everything that is not a class must be a property. Determine the domain of each property (the classes to which the property applies) and the range of each property (the kind of data value associated with the property). The default value for a property's range is "Literal".
4. Prepare an XML schema (`.xsd` file) datatype declaration for each Property range that requires a constrained data type.
5. Prepare a CDE (Common Data Element) list that includes the basic CDE annotations recommended in ISO-11179, and the basic annotation needed to describe classes, properties, and datatypes for property ranges.
6. Convert CDEs to an RDF schema.
7. Validate the RDF schema.
8. Vet the RDF schema through the intended user community.
9. Distribute the RDF schema to the public.
10. Repeat Steps 1-9 ad libitum, providing a new version number to each successive specification.

18.4. USING THE RDF SCHEMA TO SPECIFY DATA OBJECTS IN RDF

Once you have access to RDF schemas relevant to your data objects, you can write RDF documents that fully describe your data and that draw their descriptive vocabulary from one or more relevant RDF schemas.

LIST 18.4.1. GENERAL INSTRUCTIONS FOR SPECIFYING BIOMEDICAL DATA IN RDF

1. Look at your own available data for the data object. List your data as triples sequenced as: subject, metadata, and data.
2. Create an RDF header that includes the URL of each of the RDF schemas that you will use in the document. Be sure to include the Dublin Core RDF schema. This document includes the elements that describe the file itself (that is, the name of the person who created the file, the type of data contained in the file, the type of file) and is used by librarians to index your document.
3. Use the RDF schema(s) appropriate for the knowledge domain of your data object. Determine which of your listed triples have subjects that are instances of classes in the RDF schema and metadata consistent with class-appropriate properties. Type these subjects as class instances (See Chapter 11, Section 12). Check that the data values conform to any data-value constraints listed in the xsd datatype file associated with the RDF schema.
4. If you have common data elements (classes or properties) that are not part of any public RDF schema, create your own RDF schema to accommodate these elements, and use the RDF schema as the resource for those elements wherever they appear in your RDF specification document.
5. Contact the curator of the public RDF schema that is appropriate for the elements you created, and then ask if your RDF schema can be added to the public RDF schema.
6. Validate that the document is well-defined XML, syntactically correct RDF, and that all triples validate agains class-property-datatype descriptions from the RDF schema.

18.5. USE-CASE: THE LABORATORY DIGITAL IMAGING PROJECT

The Laboratory Digital Imaging Project (LDIP) is an effort sponsored by the Association for Pathology Informatics to create a data specification for histopathology images (Figure 18-1).

It is designed to convey a variety of technical, clinical, and pathologic data, along with their image binaries.

The group has prepared some tentative CDEs for their image object, and the CDEs are assigned either Class names or Property names (List 18.5.1).

Once you have your CDEs, you can write a simple Perl parsing script to convert them directly into RDF schema elements.

LIST 18.5.1. SOME CDES EXCERPTED FROM THE LDIP CDE FILE

```
Identifier:ldip:Person
Class Label:Person
versionInfo: 0.1
Registration Authority: Association for Pathology Informatics
Language:en
Datatype: Literal
comment: All persons involved in the patient's care, including
the patient him/herself, and all providers and collections of
providers.
subClassOf:Class
Contributor:Bill Moore
Date_of_contribution:05-30-2006


Identifier:ldip:Instrument
Class Label:Instrument      VersionInfo: 0.1
Registration Authority: Association for Pathology Informatics
Language:en
Datatype: Literal
Comment: All the instruments used in preparing, viewing,
and imaging a specimen. Includes: microscope, camera.
subClassOf:Class
Contributor:Bill Moore
Date_of_contribution:05-30-2006


Identifier:ldip:stain
Property Label:stain
versionInfo: 0.1
Registration Authority: Association for Pathology
   Informatics
Language:en
Datatype (required, can be regex, URL, or literal): literal
comment: For example, Unstained, Hematoxylin and eosin,
Congo Red, Masson, Periodic acid Schiff,....
domain:Reagent, Image
range: Literal
Contributor:Bill Moore
Date_of_contribution:05-30-2006
```

18.6. THE RDF SCHEMA IS PREPARED FROM A CDE LIST

A small RDF schema composed from some CDEs is shown (List 18.6.1).

LIST 18.6.1. AN ABBREVIATED RDF SCHEMA CONSTRUCTED FROM LDIP CDES

```
<?xml version="1.0" encoding="UTF-8"?>
<rdf:RDF
xmlns:rdf="http://www.w3.org/1999/02/22-rdf-syntax-ns#"
xmlns:rdfs="http://www.w3.org/2000/01/rdf-schema#"
>
<rdf:Class rdf:about="http://www.ldip.org/ldip_sch#Image">
<rdfs:label>Image</rdfs:label>
<rdfs:comment>
A digital image file, such as a jpeg, tiff, or DICOM file
</rdfs:comment>
<rdfs:subClassOf
rdfs:resource="http://www.w3.org/2000/01/rdf-schema#Class"/>
</rdf:Class>


<rdf:Property
   rdf:about="http://www.ldip.org/ldip_sch#fileName">
<rdfs:label>fileName</rdfs:label>
<rdfs:comment>
A filename can include a directory path in DOS or
UNIX notation.
</rdfs:comment>
<rdfs:domain
   rdf:resource="http://www.ldip.org/ldip_sch#report"/>
<rdfs:domain
   rdf:resource="http://www.ldip.org/ldip_sch#image"/>
<rdfs:range
   rdf:resource="http://www.ldip.org/ldip_xsd.xsd#filepath"/>
</rdf:Property>


<rdf:Property rdf:about="http://www.ldip.org/ldip_sch#lsid">
<rdfs:label>lsid</rdfs:label>
<rdfs:comment>
The life sciences identifier (lsid)
</rdfs:comment>
<rdfs:domain
   rdf:resource="http://www.ldip.org/ldip_sch#report"/>
```

```
<rdfs:domain
   rdf:resource="http://www.ldip.org/ldip_sch#image"/>
<rdfs:range
   rdf:resource="http://www.ldip.org/ldip_xsd.xsd#lsid"/>
</rdf:Property>


<rdf:Property
   rdf:about="http://www.ldip.org/ldip_sch#objective">
<rdfs:label >objective</rdfs:label>
<rdfs:comment>
A description of the microscope's objective lens.
Required elements include the lens numerical aperture, and the
magnification, both of which are floating point (real) numbers.
</rdfs:comment>      <rdfs:domain
rdf:resource="http://www.ldip.org/ldip_sch#Microscope"/>
<rdfs:range
rdf:resource="http://www.ldip.org/ldip_xsd.xsd#EnumerationObjectives"/>
</rdf:Property>


<rdf:Property rdf:about="http://www.ldip.org/ldip_sch#model">
<rdfs:label>model</rdfs:label>
<rdfs:comment>
the manufacturer's model name for an instrument
</rdfs:comment>
<rdfs:domain
rdf:resource="http://www.ldip.org/ldip_sch#Instrument"/>
<rdfs:range
rdf:resource="http://www.w3.org/2000/01/rdf-schema#Literal"/>
</rdf:Property>
<rdf:Property
   rdf:about="http://www.ldip.org/ldip_sch#serialNumber">
<rdfs:label>serial_number</rdfs:label>
<rdfs:comment>
the manufacturer's unique serial number for an instrument
</rdfs:comment>
<rdfs:domain
rdf:resource="http://www.ldip.org/ldip_sch#Instrument"/>
<rdfs:range
rdf:resource="http://www.w3.org/2000/01/ldip_xsd.xsd#serialNumber"/>
</rdf:Property>
</rdf:RDF>
```

This RDF schema lists one class (Image) and six properties (`fileName`, `lsid`, `ldipIdentifier`, `objective`, `model`, and `serialNumber`). We will use some of these in our use-case example. Let us look at the `fileName` property (List 18.6.2).

LIST 18.6.2. THE `fileName` PROPERTY

```
<rdf:Property
   rdf:about="http://www.ldip.org/ldip_sch#fileName">
<rdfs:label>fileName</rdfs:label>
<rdfs:comment>
A filename can include a directory path in DOS or UNIX
   notation.
</rdfs:comment>
<rdfs:domain
   rdf:resource="http://www.ldip.org/ldip_sch#report"/>
<rdfs:domain
   rdf:resource="http://www.ldip.org/ldip_sch#image"/>
<rdfs:range
   rdf:resource="http://www.ldip.org/ldip_xsd.xsd#filepath"/>
</rdf:Property>
```

The domain of a Property is the class of objects described by the Property. Notice that there are two domains listed for the fileName property: Report and Image. This is allowable under the rules of RDF schema. But is it necessary or wise to use two domains when one will suffice? Both Report and Image are subclasses of the father class, `Data_Object`. The easiest way to organize the RDF schema is to list `Data_Object` as the single domain for `fileName`. `Data_Object` would be the preferred domain for the `fileName` Property if all the subclasses of `Data_Object` could be described with the `fileName` Property.

18.7. REVIEW OF RDF SCHEMA PROPERTIES

LIST 18.7.1. IMPORTANT FEATURES OF THE RDF SCHEMA

1. All the information needed to prepare the RDF schema file can be included in the ISO-11179 conformant CDE list.
2. An RDF schema, like all RDF documents, is an XML file.
3. An RDF schema employs RDF syntax. It starts with the rdf description element and lists the namespaces used in the document. It then lists triples for classes and properties. Each triple begins with a resource identifier for the subject of the triple (the

name of the class or the name of the property) and is followed by simple metadata/data pairs.

4. A property can have more than one domain.
5. A property range can be a linked resource. When the property range links to a datatype declaration within an XML schema, we can constrain the range of a property.

18.8. THE JPEG IMAGES

For our use-case examples, we will use the following two public domain figures:

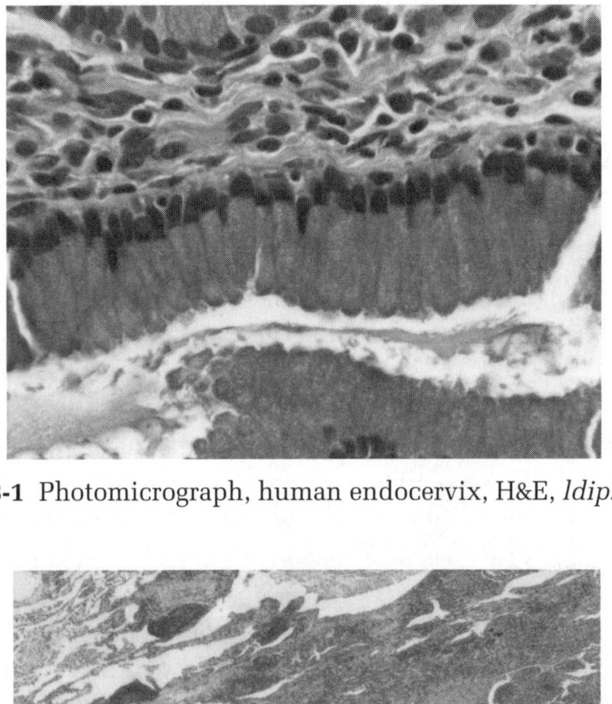

Figure 18-1 Photomicrograph, human endocervix, H&E, *ldip2103.jpg*.

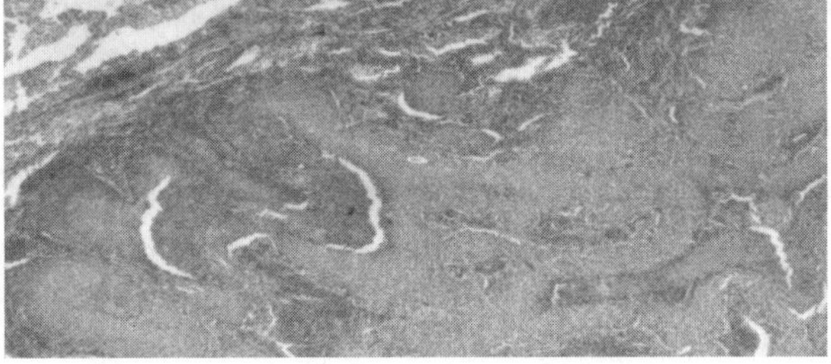

Figure 18-2 Photomicrograph, squamous cell carcinoma of human lung, H&E, *ldip2201.jpg*.

18.9. SAMPLE TEXTUAL ANNOTATION

The following is a sample annotation for a pathology photomicrograph. We will use this annotation to show how to make RDF triples from unstructured data. We will provide five different strategies for specifying an image object in conformance with an RDF schema.

> Unstructured annotation: On June 21, 2006, Bill Moore, MD, took a photo of human endocervix, using a histologic slide from an old collection of publicly distributed teaching slides of anonymous patients. He used an Olympus Model BH2 microscope, serial number 224085, under the 10x objective lens. The camera that took the digital image was an Infinity 3 CCD camera, Serial number 00169344. The produced image file, named `ldip2103`, is an H&E stained photomicrograph of normal endocervix and has the URL *http://www.gwmoore.org/ldip/ldip2103.jpg.*

18.10. FIVE OPTIONS FOR PREPARING AN RDF SPECIFICATION FOR AN IMAGE OBJECT

In the next sections, we will take the preceding structured annotation and transform it into an RDF specification for the image object. There are five ways we can do this (List 18.10.1).

LIST 18.10.1. FIVE WAYS TO SPECIFY A PHOTOMICROGRAPH IN RDF

1. RDF document points to one or more image files. The typical case would be a single RDF document, which describes one image and links to the URL where the image is stored.
2. RDF document contains an image binary converted to base64 ASCII.
3. Image file contains an RDF document inserted into the image file header.
4. Multiple RDF documents contribute to the description of an image and point to the URL where the image is stored.
5. Multiple RDF documents point to multiple images or multiple parts of a single image file.

18.11. OPTION ONE: RDF DOCUMENT POINTS TO AN IMAGE FILE

The simplest way to create an image specification is to prepare an RDF file containing triples that have Classes and Properties described in an RDF schema created for images, and to include one triple that points to the URL where the image file (the so-called image binary) is stored (List 18.11.1).

LIST 18.11.1. ONE RDF FILE DESCRIBES IMAGE AND INSTRUMENTS AND POINTS TO THE IMAGE BINARY

```
<?xml version="1.0"?>
<rdf:RDF
xmlns:rdf="http://www.w3.org/1999/02/22-rdf-syntax-ns#"
xmlns:ldip="http://www.ldip.org/ldip_sch#"
xmlns:dc="http://perl.org/dc/elements/1.1/">


<rdf:Description
   rdf:about="http://www.gwmoore.org/ldip/ldip2103.jpg">
<rdf:type rdf:resource= "http://www.ldip.org/ldip_sch#Image"/>
<dc:title>Normal endocervix</dc:title>
<dc:creator>Bill Moore</dc:creator>
<dc:date>2006-06-21</dc:date>
<ldip:instrument_id
rdf:resource="urn:www.ldip.org:ldip:Olympus_BH2_224085"/>

<ldip:instrument_id
rdf:resource="urn:www.ldip.org:ldip:Infinity_3_00169344"/>
<ldip:imageType>photomicrograph</ldip:imageType>]
<ldip:stain>H and E</ldip:stain>
<ldip:tissue>endocervix</ldip:tissue>
<ldip:organism>human</ldip:organism>
<ldip:objective>10x</ldip:objective>
<ldip:diagnosis>normal</ldip:diagnosis>
</rdf:Description>


<rdf:Description
   rdf:about="urn:www.ldip.org:ldip:Olympus_BH2_224085">
<rdf:type
   rdf:resource="http://www.ldip.org/ldip_sch#Instrument"/>
<ldip:instrumentType>Microscope</ldip:instrumentType>
<ldip:make>Olympus</ldip:make>
<ldip:model>BH2</ldip:model>
<ldip:serialNumber>224085</ldip:serialNumber>
</rdf:Description>


<rdf:Description
   rdf:about="urn:www.ldip.org:ldip:Infinity_3_00169344">
<rdf:type resource= "http://www.ldip.org/ldip_sch#Instrument"/>
```

```
<ldip:instrumentType>Camera</ldip:instrumentType>
<ldip:make>Infinity</ldip:make>
<ldip:model>3</ldip:model>
<ldip:serialNumber>00169344</ldip:serialNumber>
</rdf:Description>
</rdf:RDF>
```

18.12. OPTION TWO: RDF DOCUMENT CONTAINS IMAGE BINARY CONVERTED TO BASE64 ASCII

Though we distinguish text files from binary files, all files are actually binary files. Sequential bytes of 8 bits are converted to ASCII equivalents, and if the ASCII equivalents are alphanumerics, we call the file a text file. If the ASCII values of 8-bit sequential file chunks are nonalphanumeric, we call the files binary files. Standard format image files (such as jpeg, tiff, gif, png) are binary files. Because RDF syntax is a pure ASCII file format, image binaries cannot be directly pasted into an RDF document. However, binary files can be interconverted from Base 64ASCII using a simple software utility. Base64 ASCII files can be inserted directly into RDF documents.

18.13. CONVERTING A JPG (BINARY) IMAGE FILE TO BASE64

LIST 18.13.1. PERL SCRIPT `jpg2b64.pl` CONVERTS A JPEG FILE TO BASE64

```
#!/usr/bin/perl
use MIME::Base64::Perl;
open (TEXT,"c\:\\ftp\\ldip\\ldip2103\.jpg")||die"cannot";
#path to sample file
binmode TEXT;
$/ = undef;
$string = <TEXT>;
close TEXT;
$encoded = encode_base64($string);
open(OUT,">2103.txt");
print OUT $encoded;
close OUT;

#if you wish, the following lines will decode your Base64 files
#$decoded = decode_base64($encoded);
#open(OUT,">binary.jpg");
#binmode OUT;
#print OUT $decoded;
exit;
```

Here is an example of the same RDF document shown in the prior use-case. The only difference is that in addition to pointing to the URL that identifies the image, this document contains part of the image file converted to Base64 ASCII (List 18.13.2).

LIST 18.13.2. RDF FILE WITH BASE64 IMAGE DATA

```
<?xml version="1.0"?>
<rdf:RDF
xmlns:rdf="http://www.w3.org/1999/02/22-rdf-syntax-ns#"
xmlns:ldip="http://www.ldip.org/ldip_sch#"
xmlns:dc="http://perl.org/dc/elements/1.1/">


<rdf:Description
   rdf:about="http://www.gwmoore.org/ldip/ldip2103.jpg">
<rdf:type rdf:resource= "http://www.ldip.org/ldip_sch#Image"/>
<dc:title>Normal Endocervix</dc:title>
<dc:creator>Bill Moore</dc:creator>
<dc:date>2006-06-21</dc:date>
<ldip:instrument_id
rdf:resource="urn:www.ldip.org:ldip:Olympus_BH2_224085"/>

<ldip:instrument_id
rdf:resource="urn:www.ldip.org:ldip:Infinity_3_00169344"/>
<ldip:imageType>photomicrograph</ldip:imageType>
<ldip:stain>H and E</ldip:stain>
<ldip:tissue>endocervix</ldip:tissue>
<ldip:organism>human</ldip:organism>
<ldip:objective>10x</ldip:objective>
<ldip:diagnosis>normal</ldip:diagnosis>
<ldip:base64File>
/9j/4AAQSkZJRgABAQEAYABgAAD/2wBDAAgGBgcGBQgHBwcJCQgKDBQNDAsLDBk
SEw8UHRofHh0aHBwgJC4nICIsIxwcKDcpLDAxNDQ0Hyc5PTgyPC4zNDL/2wBDAQ
kJCQwLDBgNDRgyIRwhMjIyMjIyMjIyMjIyMjIyMjIyMjIyMjIyMjIyMjIyMjIyM
jIyMjIyMjIyMjIyMjIyMjL/wAARCAYACAADASIAAhEBAxEB/8QAHwAAAQUBAQEB
AQEAAAAAAAAAAAECAwQFBgcICQoL/8QAtRAAAgEDAwIEAwUFBAQA
.
.
.
DlphhQnOMVSxL6nRHEaWZz8tpvGBWdPpMpbO08967AW8YOcU4xIRgqKr6yl0N44
9xOSOOFOkv6Yq1a6W6HJHFdb9miznbStAhB4AzVfWl0KqY3nsjjNSuEtosE9BXC
6lrf7whT39a7jxLpOxjbah6V5leaZchySh6l04PC+196RlzrU6cFZo/9k=
</ldip:base64File>
```

```
</rdf:Description>
<rdf:Description
   rdf:about="urn:www.ldip.org:ldip:Olympus_BH2_224085">
<rdf:type
   rdf:resource="http://www.ldip.org/ldip_sch#Instrument"/>
<ldip:instrumentType>Microscope</ldip:instrumentType>
<ldip:make>Olympus</ldip:make>
<ldip:model>BH2</ldip:model>
<ldip:serialNumber>224085</ldip:serialNumber>
</rdf:Description>


<rdf:Description
   rdf:about="urn:www.ldip.org:ldip:Infinity_3_00169344">
<rdf:type resource= "http://www.ldip.org/ldip_sch#Instrument"/>
<ldip:instrumentType>Camera</ldip:instrumentType>
<ldip:make>Infinity</ldip:make>
<ldip:model>3</ldip:model>
<ldip:serialNumber>00169344</ldip:serialNumber>
</rdf:Description>
</rdf:RDF>
```

We cut the bulk of the image file from the RDF document. A complete Base64 image file can occupy hundreds or even thousands of pages of text. The Base64 rendition of `ldip2103.jpg` has a length of 508,750. It is somewhat ironic that the only way of making an image file readable within an RDF document is to create a file that is too long to actually read.

18.14. OPTION THREE: INSERTING AN RDF DOCUMENT INTO A JPEG HEADER

Almost all popular image formats contain "header" sections. The header sections contain information that is used by image-viewing software to properly display the image. Robust imaging-software applications are written with subroutines that parse through the different headers of images and extract information such as the height, width, pixel number, pixel size, pixel color, and color map index. Some headers are extensible, allowing software to insert blocks of text into the header without changing the image binary. It is easy to insert an RDF document into the header of a jpeg image file, and it is just as easy to extract the RDF triples. Here's how you do it (List 18.14.1):

LIST 18.14.1. PREPARING AN RDF DOCUMENT THAT SPECIFIES YOUR IMAGE (OR ANY OTHER DATA OBJECT)

1. Prepare your RDF document just as you would in the earlier use-case examples.
2. Use software that adds comments to the header of a jpeg file.
3. Use the new jpeg file (now with RDF comments) to display the image or to send to colleagues. When displayed, it will look exactly like the file did before the contents of the RDF document were added.
4. Use software to extract the comments from the header of the jpeg file, as needed.

The following Perl script will take the jpeg image `ldip2103.jpg` and add the RDF document from the first use-case to its header. Once created, the program extracts and displays the contents of the RDF file (List 18.14.2).

LIST 18.14.2. PERL SCRIPT `meta_jpg.pl`, TO SHOW HOW METADATA CAN BE ADDED TO A JPEG FILE

```
#!/usr/local/bin/perl
use Image::MetaData::JPEG;
my $filename = "ldip2103.jpg"; #comment:your filename here
my $file = new Image::MetaData::JPEG($filename);
die 'Error: ' . Image::MetaData::JPEG::Error() unless $file;
print "Description of JPEG file\n";
print $file->get_description();
print "\n\nRDF Annotations to JPEG file\n\n";
open (TEXT, "rdf_desc.xml")||die"cannot"; #comment:rdf document
$line = " ";
while ($line ne "")
  {
  $line = <TEXT>;
  $file->add_comment($line);
  }
unlink $filename;
$file->save($filename);
my $file = new Image::MetaData::JPEG($filename);
my @comments = $file->get_comments();
print join("",@comments);
exit;
```

This Perl script requires the freely available open-source module `Image::MetaData::JPEG` located at http://www.cpan.org/.

18.15. OPTION FOUR: SPECIFYING AN IMAGE WITH MULTIPLE RDF FILES

There are times when the annotative data (that is, the non-binary data) for a data object's specification must be distributed in multiple files.

This is one of the most important reasons for using a data specification, rather than a data standard. The specification permits you to create a dynamic object composed of informational pieces that can be updated, so that the content and value of a specified image object increases over time. A data standard obligates you to compose a static data file. If the standard data file contains information that cannot be shared (due to human subject risks, or to intellectual property encumbrances), the standard file usually cannot be distributed.

A specification may consist of multiple files connected by URL pointers. If component files contain privileged information, the data object's specification can be distributed with access restricted to specified files (Lists 18.15.1 and 18.15.2).

LIST 18.15.1. FILE 1 POINTS TO IMAGE BINARY AND TO FILES 2 AND 3.

```
<?xml version="1.0"?>
<rdf:RDF
xmlns:rdf="http://www.w3.org/1999/02/22-rdf-syntax-ns#"
xmlns:ldip="http://www.ldip.org/ldip_sch#"
xmlns:dc="http://perl.org/dc/elements/1.1/">

<rdf:Description
   rdf:about="http://www.gwmoore.org/ldip/ldip2103.jpg">
<rdf:type rdf:resource= "http://www.ldip.org/ldip_sch#Image"/>
<dc:title>Normal Small Intestine</dc:title>
<dc:creator>Bill Moore</dc:creator>
<dc:date>2006-06-21</dc:date>
<ldip:instrument_id
rdf:resource="urn:www.ldip.org:ldip:Olympus_BH2_224085"/>
<ldip:linkedFile rdf:resource="http://www.ldip.org/file2"/>
<ldip:instrument_id
rdf:resource="urn:www.ldip.org:ldip:Infinity_3_00169344"/>
<ldip:linkedFile rdf:resource="http://www.ldip.org/file3"/>
<ldip:imageType>photomicrograph</ldip:imageType>
<ldip:stain>H and E</ldip:stain>
```

```
<ldip:tissue>endocervix</ldip:tissue>
<ldip:organism>human</ldip:organism>
<ldip:objective>10x</ldip:objective>
<ldip:diagnosis>normal</ldip:diagnosis>
</rdf:Description>
</rdf:RDF>
```

LIST 18.15.2. FILE 2 OF MULTI-FILE IMAGE DESCRIBES A MICROSCOPE REFERENCED BY FILE 1

```
<?xml version="1.0"?>
<rdf:RDF
xmlns:rdf="http://www.w3.org/1999/02/22-rdf-syntax-ns#"
xmlns:ldip="http://www.ldip.org/ldip_sch#"
xmlns:dc="http://perl.org/dc/elements/1.1/">
<rdf:Description
   rdf:about="urn:www.ldip.org:ldip:Olympus_BH2_224085">
<rdf:type
   rdf:resource="http://www.ldip.org/ldip_sch#Instrument"/>
<ldip:instrumentType>Microscope</ldip:instrumentType>
<ldip:make>Olympus</ldip:make>
<ldip:model>BH2</ldip:model>
<ldip:serialNumber>224085</ldip:serialNumber>
</rdf:Description>
</rdf:RDF>
```

LIST 18.15.3. RDF FILE 3 OF MULTI-FILE IMAGE DESCRIBES A CAMERA REFERENCED BY FILE 1

```
<?xml version="1.0"?>
<rdf:RDF
xmlns:rdf="http://www.w3.org/1999/02/22-rdf-syntax-ns#"
xmlns:ldip="http://www.ldip.org/ldip_sch#"
xmlns:dc="http://perl.org/dc/elements/1.1/">
<rdf:Description
   rdf:about="urn:www.ldip.org:ldip:Infinity_3_00169344">
<rdf:type resource= "http://www.ldip.org/ldip_sch#Instrument"/>
<ldip:instrumentType>Camera</ldip:instrumentType>
<ldip:make>Infinity</ldip:make>
<ldip:model>3</ldip:model>
<ldip:serialNumber>00169344</ldip:serialNumber>
</rdf:Description>
</rdf:RDF>
```

18.16. OPTION FIVE: SPECIFYING MULTIPLE IMAGE FILES AND MULTIPLE RDF DOCUMENTS

Multiple RDF documents may point to multiple images or multiple parts of a single image file. Suppose, as in the previous example, that the triples relevant to your image lie in multiple RDF files. Suppose further that your image is just one of a set of images that were all obtained during the same session, and that all the images apply to the same patient. This situation is routine for radiologic images, wherein dozens of images transecting the brain or the abdomen might form part of the same report.

How might you annotate this complex set of data files and image binaries? Simply include an RDF assertion for each image (Lists 18.16.1, 18,16.2, and 18.16.3).

LIST 18.16.1. FILE 1 DESCRIBES TWO IMAGES AND POINTS TO FILES 2 AND 3

```
<?xml version="1.0"?>
<rdf:RDF
xmlns:rdf="http://www.w3.org/1999/02/22-rdf-syntax-ns#"
xmlns:ldip="http://www.ldip.org/ldip_sch#"
xmlns:dc="http://perl.org/dc/elements/1.1/">


<rdf:Description
   rdf:about="http://www.gwmoore.org/ldip/ldip2103.jpg">
<rdf:type rdf:resource= "http://www.ldip.org/ldip_sch#Image"/>
<dc:title>Normal Endocervix</dc:title>
<dc:creator>Bill Moore</dc:creator>
<dc:date>2006-06-21</dc:date>
<ldip:instrument_id
rdf:resource="urn:www.ldip.org:ldip:Olympus_BH2_224085"/>
<ldip:linkedFile rdf:resource="http://www.ldip.org/file2">

<ldip:instrument_id
rdf:resource="urn:www.ldip.org:ldip:Infinity_3_00169344"/>
<ldip:linkedFile rdf:resource="http://www.ldip.org/file3"/>
<ldip:imageType>photomicrograph</ldip:imageType>
<ldip:stain>H and E</ldip:stain>
<ldip:tissue>small intestine</ldip:tissue>
<ldip:organism>human</ldip:organism>
<ldip:objective>10x</ldip:objective>
```

```
<ldip:diagnosis>normal</ldip:diagnosis>
</rdf:Description>
<rdf:Description
   rdf:about="http://www.gwmoore.org/ldip/ldip2201.jpg">
<rdf:type rdf:resource= "http://www.ldip.org/ldip_sch#Image"/>
<dc:title>Squamous Carcinoma</dc:title>
<dc:creator>Bill Moore</dc:creator>
<dc:date>2006-06-21</dc:date>
<ldip:instrument_id
rdf:resource="urn:www.ldip.org:ldip:Olympus_BH2_224085"/>
<ldip:linkedFile rdf:resource="http://www.ldip.org/file2">

<ldip:instrument_id
rdf:resource="urn:www.ldip.org:ldip:Infinity_3_00169344"/>
<ldip:linkedFile rdf:resource="http://www.ldip.org/file3"/>
<ldip:imageType>photomicrograph</ldip:imageType>
<ldip:stain>H and E</ldip:stain>
<ldip:tissue>lung</ldip:tissue>
<ldip:organism>human</ldip:organism>
<ldip:objective>2.5x</ldip:objective>
<ldip:diagnosis>squamous cell carcinoma</ldip:diagnosis>
</rdf:Description>
</rdf:RDF>
```

LIST 18.16.2. FILE 2 DESCRIBES A MICROSCOPE THAT VIEWED THE TWO IMAGES DESCRIBED IN FILE 1

```
<?xml version="1.0"?>
<rdf:RDF
xmlns:rdf="http://www.w3.org/1999/02/22-rdf-syntax-ns#"
xmlns:ldip="http://www.ldip.org/ldip_sch#"
xmlns:dc="http://perl.org/dc/elements/1.1/">
<rdf:Description
   rdf:about="urn:www.ldip.org:ldip:Olympus_BH2_224085">
<rdf:type
   rdf:resource="http://www.ldip.org/ldip_sch#Instrument"/>
<ldip:instrumentType>Microscope</ldip:instrumentType>
<ldip:make>Olympus</ldip:make>
<ldip:model>BH2</ldip:model>
<ldip:serialNumber>224085</ldip:serialNumber>
</rdf:Description>
</rdf:RDF>
```

LIST 18.16.3. FILE 3 DESCRIBES A CAMERA THAT TOOK THE TWO IMAGES DESCRIBED IN FILE 1

```
<?xml version="1.0"?>
<rdf:RDF
xmlns:rdf="http://www.w3.org/1999/02/22-rdf-syntax-ns#"
xmlns:ldip="http://www.ldip.org/ldip_sch#"
xmlns:dc="http://perl.org/dc/elements/1.1/">
<rdf:Description
   rdf:about="urn:www.ldip.org:ldip:Infinity_3_00169344">
<rdf:type resource= "http://www.ldip.org/ldip_sch#Instrument"/>
<ldip:instrumentType>Camera</ldip:instrumentType>
<ldip:make>Infinity</ldip:make>
<ldip:model>3</ldip:model>
<ldip:serialNumber>00169344</ldip:serialNumber>
</rdf:Description>
</rdf:RDF>
```

The same approach can be used to reference multiple images within a single image file. The `rdf:about` attribute can point to any file block or file element that contains the part of the image that is the intended subject of the triples (for example, region of interest, thumbnail, tile, waveform, color map).

18.17. PORTING BETWEEN DATA SPECIFICATIONS AND DATA STANDARDS

Because all the data in a specification is fully described, it is easy to write software that will port a specification into a data standard. To have full compatibility between a data specification and data standard, the specification must contain all of the required data elements of the data standard (List 18.17.1).

LIST 18.17.1. PORTING BETWEEN A DATA STANDARD AND A DATA SPECIFICATION

1. Study the data standard and write an RDF schema (or supplement an existing RDF schema) with classes and properties appropriate for the data standard.
2. Write software that will parse the RDF document in which the data object is specified (always trivial) and transform the triples into a document that conforms to the data standard (sometimes trivial).

Because Perl's strongest feature is data transformation, it is the ideal language for porting between an RDF data specification and any chosen data standard.

18.18. SUMMARY

This book is focused on using Perl to solve common computational tasks in the fields of biology and medicine. The most pervasive task in biology and medicine is organizing and sharing data. RDF, or something very similar to RDF, will likely become the most important method for sharing biomedical data resources. Perl can create, transform, locate, parse and interrogate RDF documents.

CHAPTER

19

Use-Case: Data Mining OMIM (Online Mendelian Inheritance in Man)

19.1. BACKGROUND

Much of the extant literature in the field of data mining reminds me of the instructions for a foolproof fly-killing kit (List 19.1.1).

LIST 19.1.1. FOOLPROOF METHOD TO KILL FLIES

1. Place fly on smooth, hard surface.
2. Obtain stout, wooden board.
3. Smack fly hard with wooden board.

It is never easy to place a fly on a smooth, hard surface. Likewise, it is never easy to place biomedical data into a data mining application. It is silly to think you can buy software that will satisfy your data mining aspirations.

The purpose of this chapter is to provide examples of the kinds of data mining questions that a medically-oriented biologist might ask and to provide a step-wise approach to answering those questions, using Perl.

19.2. THE HYPOTHESES

Some scientific questions cannot be answered by experience, by reviewing the scientific literature, by conducting experiments, by thinking creatively about a problem, or by constructing a computational model. Some problems will be answered only by having a complete set of information for a knowledge domain, and by having methods to interrogate the knowledge contained in that domain.

For instance, if you wanted to know the total number of different patients served by a hospital, you would need access to a listing of all patients, and you would need a method to count those patients that ignores replicate entries for the same patient. There is simply no way to arrive at an answer without the list of patients. Having a partial list of patients from every hospital clinic or a complete list of patients in a single clinic will not suffice. You might be able to estimate the total number (by multiplying the number of patients treated in a clinic by the number of clinics in the hospital), but you would need some method to eliminate replicate visits (from patients who visit many clinics). Even then, the estimate might be far off the mark.

Much of what we think we know about medicine is based on the broad professional experiences of generations of doctors. However, broad experience cannot help in areas where no single physician can expect to encounter all of the experiences relevant to a field. An example is found in the realm of inherited diseases of humans. There are thousands of inherited diseases. Some of these diseases are rare, with only a few hundred probands worldwide. A clinical geneticist cannot acquire professional experience with every genetic disease known to mankind.

The only way to answer a question that might apply to the class of all inherited diseases is to start with a listing of every genetic disease.

One of the greatest benefits of modern biomedicine is the availability of comprehensive datasets, that is, datasets that contain all or nearly all of the items within a knowledge domain (List 19.2.1).

Several comprehensive biomedical data collections are listed in the yearly database issue of the journal *Nucleic Acids* (69).

My own interests lie within the field of cancer research. There are abundant statistics available on the occurrence of cancer in the general population of the United States and of the world. The most common tumors of humans are the so-called epithelial tumors, or carcinomas. Carcinomas

LIST 19.2.1. SOME COMPREHENSIVE DATASETS FREELY AVAILABLE TO BIOMEDICAL DATA MINERS

- OMIM, Online Mendelian Inheritance in Man, containing all of the genetic diseases of man with detailed descriptions of clinical features and pathobiology
- Human Genome Project, containing the DNA sequence of the human genome
- GO, an ontology of nearly every studied human gene
- Taxonomy for the Developmental Lineage of Neoplasms, containing over 100,000 classified names of nearly every human neoplasm

usually come in the form of gland-forming tumors (adenocarcinomas) or surface-forming tumors (squamous cell carcinomas composed of flat, pavement-like cells).

Epithelial tumors account for about 90 percent of the tumors that occur in humans. How do we know this? There are thousands of different types of tumors, but the five tumors (List 19.2.2) that account for more than half of all malignant tumors found in humans are epithelial tumors (70).

The U.S. government's SEER (Surveillance, Epidemiology, and End Results) program does not collect data for the two most common epithelial cancers of humans: squamous cell carcinoma of the skin and basal cell carcinoma of the skin (71). In 1990, 600,000 new cases of basal cell cancer or squamous cell cancer were diagnosed in the United States. These two tumors are so common that the expense to register their incidence would be prohibitive. The incidence of these two epithelial tumors not included in most cancer-incidence registries is of a magnitude roughly equivalent to the incidence of all other tumors combined. If these tumors were taken into account, epithelial tumors would account for more than 95% of the tumors of humans.

Most non-epithelial tumors are sarcomas. Sarcomas include tumors composed of spindle-shaped cells that derive from the connective tissues (fibrous tissue, muscle, bone, cartilage, joints, vessels). Sarcomas arising from the soft tissues (fibrous tissue muscle, joints, vessels) account for fewer than 1 percent of all cancers. Hard tissue sarcomas (arising from bone and cartilage) are much less common than soft tissue tumors, accounting for about 2,500 new cases each year in the United States (72). Lymphomas and leukemias are sometimes grouped with the sarcomas because they arise from mesenchymal cells, the same cell lineage that includes the hard and soft tissues. The lymphomas and leukemias account for fewer than 5 percent of all cancers (70). Leukemias account for fewer cancers than lymphomas (70). Combined, sarcomas of soft tissues, hard tissues, lymphomas, and leukemias account for less than 10 percent of cancers.

LIST 19.2.2. TOP-RANKED GLOBAL CANCER INCIDENCE FOR BOTH GENDERS, 2000 (70)

	Percentage of Total Cancer Burden
Trachea, bronchus, and lung	12.7
Stomach	10.2
Liver	10.0
Colon and rectum	9.5
Breast	6.1
Oesophagus	5.3

What accounts for this difference in incidence between epithelial tumors and sarcomas? Is there some genetic secret that makes one kind of tissue more susceptible than another? Or is the problem environmental? Are epithelial cells exposed to more carcinogens than the mesenchymal cells that give rise to sarcomas? Are epithelial cells more sensitive to carcinogens?

My opinion is that because epithelial cells line the outer (skin) and inner (alimentary tract) surfaces of the body, those cells are directly exposed to carcinogens that bathe the surfaces of the human body (via sun, air, and food). Mesenchymal cells, which are sandwiched between epithelial surfaces, are not directly exposed to carcinogens. In addition, epithelial cells tend to divide often (increasing their susceptibility to mutational agents) and tend to activate carcinogens more than do mesenchymal cells. Together, these factors might account for the greater incidence of epithelial tumors (11).

If my hypothesis were valid, we might expect that the differences in incidence of epithelial tumors and sarcomas might vanish if we confine our attention to tumors that occur as part of an inherited genetic condition. Examples of genetic conditions associated with tumors are Lynch syndrome and Von-Hippel Lindau syndrome. The rationale here is that if a tumor arose from a familial germline mutation, then the conditions that favor the occurrence of epithelial tumors over sarcomas (that is, post-natal exposure and sensitivity to environmenetal carcinogens) might not apply.

19.3. THE SPECIFIC QUESTIONS

LIST 19.3.1. HOW CAN WE USE PERL TO ANSWER THE FOLLOWING QUESTIONS?

1. What fraction of the inherited conditions of humans are associated with a predisposition to develop cancer?
2. Are the types of cancers occurring in inherited conditions different from the types of cancers occurring in the general population?
3. When an inherited condition is associated with more than one type of cancer, do the different types of cancers have anything in common with each other?
4. Does the occurrence of cancer in inherited diseases tell us anything about the biological properties of the cancer?
5. Do lessons learned about the cancers occurring in inherited conditions provide any information that might be useful in the treatment of cancers?

The questions in List 19.3.1 cannot be answered from personal experience, and they cannot be answered by an over-the-counter software application. These questions cannot be translated into a simple database query. They can be answered only by writing a data-mining program fitted to the problem at hand.

19.4. THE DATA

To answer any of these questions, we need a complete listing of all the inherited conditions of humans. For each condition, a complete description of all the lesions and medical conditions that occur as part of the condition is needed. At the time of this writing, the most complete, extensively described resource listing all the inherited conditions of humans is the publicly available Online Mendelian Inheritance in Man database (OMIM). This resource contains information on over 16,000 conditions, and the entire dataset can be downloaded from a public ftp site (see Appendix). The OMIM database is contained in a single file exceeding 112 megabytes. Many genetically inherited conditions are exceedingly rare, sometimes occurring exclusively in particular geographic locations. No individual geneticist or medical institution has sufficient experience with genetic diseases to reach convincing generalizations for the full range of genetic conditions. By parsing data on every condition, it becomes possible to reach general conclusions that apply to classes and subclasses of genetically inherited diseases.

19.5. APPROACH

Data analysis of OMIM can be difficult because the text that accompanies each OMIM entry is narrative (that is, free-text prose including extracted findings from a variety of articles from the medical literature, along with the citations for each article). This means that a tumor of the liver might have been listed as a liver adenocarcinoma in one OMIM entry, as a hepatocellular carcinoma in another entry, or as a hepatoma elsewhere.

We will use a neoplasm nomenclature, the Developmental Lineage Classification and Taxonomy of Neoplasms, which was constructed to aggregate synonymous neoplasm terms (see Appendix). This nomenclature contains about 145,000 terms and should capture almost all occurrences of neoplasm terms contained in OMIM. Because the nomenclature assigns a single code to all synonymous neoplasm terms, we can collect all the genetic conditions associated with specific neoplasms regardless of the neoplasm terminology used in the OMIM records.

19.6. FUNCTIONS OF THE PERL SCRIPT

We will need a Perl script that parses the OMIM file (List 19.6.1).

LIST 19.6.1. FUNCTIONS OF THE PERL SCRIPT `omim_4.pl`

1. Counts the total number of OMIM entries
2. Counts the number of OMIM entries that contain a reference to a neoplasm and extracts a file of these entries that can be used to assess the specificity of the script
3. Counts the tumors contained in OMIM and classifies them by histologic lineage (to distinguish epithelial tumors from sarcomatous tumors)
4. Counts those genetic syndromes that predispose to tumors deriving from more than one developmental lineage (that is, syndromes that predispose to both sarcomas and to epithelial tumors) and extracts a file of these OMIM records
5. Prepares a file that summarizes the tumor lineages represented in all those records for which multiple lineages of cancers are encountered
6. Determines the time required to execute the script (so that our productivity can be compared with workers who do not use Perl)

All of the functions in List 19.6.1 are accomplished with the `omim_4.pl` script.

19.7. HASHING NEOPLASM TERMS

`Omim_4.pl` is a Perl script that exceeds 250 lines of code. The full script is too long to include in this book, but is freely available from the Jones and Bartlett Web site.

Let us look at a few snippets from the complete script to see how it operates (List 19.7.1).

This snippet prepares two hashes, `%literalhash` and `%doubhash`, from the Developmental Lineage Classification and Taxonomy of Neoplasms (see Appendix) (11). Any nomenclature containing a comprehensive listing of neoplasm terms will suffice, but the Neoplasm Classification is far and away the largest and most complete collection of classified neoplasm terms.

The Neoplasm Classification lists entries as shown in the following lines:

LIST 19.7.1. PERL SNIPPET FROM `omim_4.pl` **PREPARES HASH OF NEOPLASM TERMS AND DOUBLETS FROM NEOPLASM NOMENCLATURE**

```
$line =~ /\"\> ?(.+) ?\<\//;
$phrase = $1;
$phrase =~ s/\'s//g;
$phrase =~ s/\,/ /g;
$phrase =~ s/[^a-z0-9 \-]/ /g;
$phrase =~ s/\b([a-z]+oma)s/$1/g;
$phrase =~ s/\b(tumo[u]?r)s/$1/g;
$phrase =~ s/\b(neoplasm)s/$1/g;
$phrase =~ s/\b(leuk[a]?emia)s/$1/g;
$phrase =~ s/ +/ /g;
if ($phrase !~ / /)
   {
   $literalhash{$phrase} = $code;
   next;
   }
$literalhash{$phrase} = $code;
@hoparray = split(/ /,$phrase);
for ($i=0;$i<(scalar(@hoparray)-1);$i++)
    {
    $doublet = "$hoparray[$i] $hoparray[$i+1]";
    next unless ($doublet =~ /[a-z]+[a-z0-9\-]*
    [a-z]+[a-z0-9\-]*/);
    $doubhash{$doublet}="";
    }
```

```
<name code = “C7631000”>malignant mesothelioma
arising in pericardium</name>
<name code = “C7631000”>malignant mesothelioma
involving pericardium</name>
```

It is easy to parse through the file, extracting phrases with a simple `regex` operation:

```
$line =~ /\”\> ?(.+) ?\<\//;
$phrase = $1;
```

The hash `%literalhash` assigns the term phrases as keys and the corresponding codes as values:

```
$literalhash{$phrase} = $code;
```

The hash %doubhash assigns word doublets, extracted from the collection of term phrases, as keys. The values for every key are assigned the empty string:

```
@hoparray = split(/ /,$phrase);
for ($i=0;$i<(scalar(@hoparray)-1);$i++)
     {
     $doublet = "$hoparray[$i] $hoparray[$i+1]";
     #hops through phrases, two words at a time
     next unless ($doublet =~ /[a-z]+[a-z0-9\-]*
     [a-z]+[a-z0-9\-]*/);
     $doubhash{$doublet}="";
     }
```

Once the doublet array has been created, we can parse through the doublets of the OMIM text, concatenating doublets that match doublets from the Neoplasm Classification and extracting concatenated strings that match full neoplasm terms (that is, concatenations that exist in %literalhash). This section of the script was described in Chapter 5, Section 2.

19.8. ASSIGNING LINEAGE TO MATCHED NEOPLASM TERMS

The tumors in the Neoplasm Taxonomy are assigned ancestral lineages in the Classification, based on their presumed histogenetic origin. For instance, an adenocarcinoma of the colon is a tumor of endodermal origin. A squamous carcinoma of the skin is a tumor of ectodermal origin. A fibroscarcoma is a tumor of mesodermal origin. Every tumor in the taxonomy (which includes about 145,000 terms) is assigned a place in the Classification, and every tumor has an ancestry of classes. This is analogous to the familiar phylogentic classification of animals. The Neoself file is prepared from the Neoplasm Classification and lists each tumor, its assigned synonymy code, and its complete ancestry (List 19.8.1).

In our script, we want to assign each tumor found in OMIM to its ancestral lineage. In general, epithelial tumors have an ancestral lineage that derives from endoderm or ectoderm. Sarcomas have an ancestral lineage that derives from mesoderm. Once we have this information, we can see if the tumors occurring in inherited conditions tend to have an ancestral lineage different from the tumors occurring in the general population (List 19.8.2 and 19.8.3).

LIST 19.8.1. SAMPLE OF FOUR ANCESTRY RECORDS IN THE `Neoself` FILE

```
#1|teratoma|C3403000|totipotent_or_multipotent_differentiating>
   primitive_differentiating>primitive>embryonic>neoplasms>
   tumor_classification>
#2|embryonal
   ca|C3752000|totipotent_or_multipotent_differentiating>
   primitive_differentiating>primitive>embryonic>neoplasms>
   tumor_classification>
#3|embryonal
   cancer|C3752000|totipotent_or_multipotent_differentiating>
   primitive_differentiating>primitive>embryonic>neoplasms>
   tumor_classification>
#4|embryonal carcinoma|C3752000|
   totipotent_or_multipotent_differentiating>
   primitive_differentiating>primitive>embryonic>
   neoplasms>tumor_classification>
```

LIST 19.8.2. PERL SNIPPET FROM `omim_4.pl` ASSIGNS DEVELOPMENTAL LINEAGE TO OMIM TUMORS

```
open (TEXT, neoself)||die"Can't open file";
$line = " ";
%neoself;
my @types = ("primitive", "endoderm_or_ectoderm",
   "mesoderm", "neuroectoderm");
while ($line ne "")
  {
  $line = <TEXT>;
  next if $line !~ /^[0-9]+\|/;
  my @linearray = split(/\|/,$line);
  foreach $thing (@types)
    {
    next if ($linearray[3] !~ /$thing/);
    $neoself{$linearray[2]} = $thing;
    }
  }
close TEXT;
```

LIST 19.8.3. THE MAJOR FAMILIES OF TUMORS IN THE NEOPLASM CLASSIFICATION

1. primitive
2. endoderm_or_ectoderm
3. mesoderm
4. neuroectoderm

19.9. COUNTING AND CLASSIFYING THE OMIM RECORDS CONTAINING NEOPLASM TERMS

The following section of Perl code collects the matched neoplasm terms in OMIM records, assigns the family in which each term belongs, and determines if there is lineage infidelity in the types of tumors that occur in the syndrome. Lineage infidelity, in this context, applies when a genetic syndrome produces tumors from two different lineage families (for example, tumors from endoderm and tumors from mesoderm). We will use this information later in our analysis of the data (List 19.9.1).

LIST 19.9.1. PERL SNIPPET FROM `omim_4.pl` COUNTS SYNDROMES WITH TUMORS OF MORE THAN ONE LINEAGE

```
while ((my $key, my $value) = each(%subhash))
    {
    print OUT "          \<v\:autocode term\=\"$key\"
    code\=\"$value\" type\=\"$neoself{$value}\" \/\>\n";
    if (exists $neoself{$value})
        {
        my $type = $neoself{$value};
        $valuecount{$type}++;
        $smallvaluecount{$type}++;
        }
    }
if (scalar(keys(%smallvaluecount)) == 2)
    {
    if (exists $smallvaluecount{"primitive"})
        {
        undef %smallvaluecount;
        next;
        }
```

```
                print BAD "\n";
                print BAD $hold;
                foreach $letter (sort (keys(%smallvaluecount)))
                    {
                    print TWO "$letter ";
                    }
                print TWO "\n";
                while ((my $key, my $value) = each(%subhash))
                   {
                   print BAD "    $key $value $neoself{$value}\n";
                   }
                while (($key, $value) = each(%smallvaluecount))
                    {
                    print BAD "    $key => $value\n";
                    }
                }
             undef %smallvaluecount;
             undef %subhash;
        print OUT"      \<\/rdf\:Description\>\n";
```

19.10. EXAMINING THE RESULTS

The results from the omim_4.pl script that appear on the monitor are shown in List 19.10.1. In addition to these summary results, omim_4.pl prepares several external files, OMIM.XML, OMIM.BAD, and OMIM.TWO.

LIST 19.10.1. SCREEN OUTPUT OF omim_4.pl

```
c:\ftp>perl omim_4.pl
The time to code was 112 seconds
The number of records in omim is 17770
The number of tumorous records in omim is 1047
The number of primitive tumors listed in omim is 359
The number of endoderm_or_ectoderm tumors listed in omim is 1299
The number of mesoderm tumors listed in omim is 1864
The number of neuroectoderm tumors listed in omim is 762
```

The OMIM.XML file contains all the extracted records of OMIM files that contained terms that matched the Neoplasm Classification.

OMIM.BAD contains the individual records of syndromes that were associated with tumors of two lineage families (List 19.10.2).

LIST 19.10.2. AN EXAMPLE OF A SINGLE RECORD IN `OMIM.BAD`

```
  <rdf:Description about="urn:OMIM-145000">
  <dc:title>#145000 hyperparathyroidism 1; hrpt1
    ;;hyperparathyroidism,          familial isolated primary;
    fihp parathyroid adenoma,          familial, included
   </dc:title>
 pituitary adenoma C3329000 endoderm_or_ectoderm
 fibroma C3041000 mesoderm
 parathyroid adenoma C3916000 endoderm_or_ectoderm
 carcinoid C4139100 endoderm_or_ectoderm
 parathyroid carcinoma C4906000 endoderm_or_ectoderm
 carcinoid of the thymus C6430000 endoderm_or_ectoderm
 parathyroid tumor C3313000 endoderm_or_ectoderm
 carcinoid tumor C4139100 endoderm_or_ectoderm
 endoderm_or_ectoderm => 7
 mesoderm => 1
```

`OMIM.TWO` is a simplified file that makes it easy to count the syndromes of two-lineage infidelity (List 19.10.3).

LIST 19.10.3. NEOPLASM SYNDROMES WITH TWO-LINEAGE TUMOR INFIDELITY

```
ectoderm/mesoderm            72 omim records
ectoderm/neuroectoderm       27 omim records
mesoderm/neuroectoderm       40 omim records
Total                        139 omim records
```

19.11. DISCUSSION

The full discussion of the results of the `omim_4.pl` script lies outside the realm of this book and would be of interest to only a small subset of cancer researchers.

Suffice it to say that about 6 percent (1,047/17,770) of OMIM syndromes are associated with the names of neoplasms. Of these neoplasms, the mesodermal tumors (composed primarily of sarcomas) represent the largest number, with 1,864 OMIM records mentioning one or more mesodermal neoplasms. Tumors of endodermal or ectodermal origin (the epithelial tumors) account for 1,299 named neoplasm references.

Of particular interest to my own research are those OMIM records associated with more than one neoplasm and for which the neoplasms associated with the syndrome come from two different developmental lineages. In my opinion, genetic changes associated with tumors of multiple developmental lineages might share certain properties not found with genetic changes that lead to tumors of a single developmental lineage. A biological question, therefore, spurred me to write the Perl script to extract OMIM records with tumor lineage discordances.

Perhaps the most important output of the Perl script are the files containing the extracted OMIM records corresponding to every record containing the name of a neoplasm and to every record containing neoplasms with a lineage discordance. Only by reviewing the records in these files can I determine that the different category counts were accurate. In point of fact, a quick review of the lineage-discordance records is sufficient to show that many sources of discordance are lesions that occur so frequently in the general population or so infrequently among the probands that those lesions might not deserve to be included as part of the genetic syndrome. In any event, data miners must understand that the data collected from a Perl script must be carefully reviewed along with the extracted records, before conclusions are published.

We began our study with a simple hypothesis (List 19.11.1).

LIST 19.11.1. ORIGINAL HYPOTHESIS FOR OUR STUDY

1. Epithelial tumors occur with high frequency in the general population because epithelial cells are better targets for environmental carcinogens than are stromal (non-epithelial cells).
2. When all cells in the body are created with genetic changes that predispose cancer (that is, in inherited cancer syndromes) we would expect a similar distribution of tumors from every developmental lineage (that is, epithelial tumors should not account for most inherited tumors).

Our preliminary observation is consistent with the long-held perception that sarcomas are frequently associated with inherited cancer syndromes. This observation starkly contrasts with the low incidence of sarcomas in the general population.

19.12. A GENERAL APPROACH TO DATA MINING

Data mining is a non-trivial pursuit (List 19.12.1).

Every aspect of a data-mining operation requires a high level of expertise. Accessing large datasets often requires an understanding of network protocols (for example, Web services) and network security. Data

LIST 19.12.1. TYPICAL REQUIREMENTS OF A MODERN DATA-MINING EFFORT

- Access to multiple large datasets
- Software to normalize, transform, or harmonize data among different datasets, so the data can be sensibly compared
- Nomenclatures and classifications for grouping synonymous data and relating data groups by shared properties
- Software to conduct computationally meaningful operations on the datasets

obtained from multiple databases is often heterogeneous and requires sophisticated data-organization skills (for example, RDF). The terms used to represent data coming from different sources often must be coded by a unifying nomenclature. This requires an understanding of the knowledge domain. Often, relationships between different types of data are important to the data miner. Using these data relationships requires a deep understanding of the semantic and computational roles of ontologies.

The complex steps in data mining make it impossible to design a universal data-mining software application. The flexibility of scripting languages, such as Perl, provides a more realistic solution.

References (Commented)

1. Zipf's law. Wikipedia, the free encyclopedia Web site: http://en.wikipedia.org/wiki/Zipf's_law. Comment. Excellent historical and technical discussion of the Zipf distribution.
2. Berman, J. J. *Pathology Abbreviated: A Long Review of Short Terms.* Archives of Pathology and Laboratory Medicine 128:347–352, 2004. Comment. Classifying abbreviations supports a logical approach to the development of class-specific algorithms designed to expand abbreviations.
3. Mitchell, K. J., M. J. Becich, J. J. Berman, W. W. Chapman, J. Gilbertson, D. Gupta, J. Harrison, E. Legowski, and R. S. Crowley. "Implementation and evaluation of a negation tagger in a pipeline-based system for information extract from pathology reports." *Medinfo* 11(Pt 1):663–667, 2004. Comment. Autocoders need a way to distinguish positive assertions (for example, tumor is present) from negative assertions (for example, tumor is not present, no tumor is seen). Otherwise, autocoding software miscounts the number of reported diseases. Negation extractors are used to distinguish occurring diagnostic entities from negative assertions that contain diagnostic terms.
4. Kim, W. and W. J. Wilbur. "Corpus-based statistical screening for phrase identification." *J Am Med Inform Assoc* 7:499–511, 2000. Comment. Everyone appreciates a good index section in a textbook. In this thoughtful article, six methods for automatically selecting candidate index phrases from text are described and tested.
5. Burrows, M. and D. J. Wheeler. "A Block-sorting Lossless Data Compression Algorithm." SRC Research Report 124, May 10, 1994. Comment. The original paper describing the Burrows-Wheeler transform (BWT).

6. BZIP2. http://www.bzip.org/. Comment. A freely available popular software implementation of a compression algorithm based on the BWT transform.
7. Sadakane, K. "New text indexing functionalities of the compressed suffix array." *J of Algorithms* 48:294–313, 2003. Comment. Suffix arrays are closely related to the BWT. This paper provides another example of the power of suffix-based indexes to organize and retrieve information.
8. Healy J., E. E. Thomas, J. T. Schwartz, and M. Wigler. "Annotating large genomes with exact word matches." Genome Research 13:2306–2315, 2003. http://www.genome.org/cgi/content/full/13/10/2306. Comment. Very clever use of the BWT to index the human genome.
9. Berman, J. J. "Modern classification of neoplasms: reconciling differences between morphologic and molecular approaches." *BMC Cancer* 5:100, 2005. http://www.biomedcentral.com/1471-2407/5/100. Comment. A new classification can sometimes resolve certain questions that arise in alternate classifications.
10. Berman, J. J. "Tumor taxonomy for the developmental lineage classification of neoplasms." *BMC Cancer* 4:88, 2004. http://www.biomedcentral.com/1471-2407/4/88/abstract. Comment. A taxonomy is the list of domain instances that fill a classification. This paper describes the taxonomy for the Developmental Lineage Classification of Neoplasms.
11. Berman, J. J. "Tumor classification: molecular analysis meets Aristotle." *BMC Cancer* 4:10, 2004. http://www.biomedcentral.com/1471-2407/4/10. Comment. Neoplasms can be classified by their developmental lineage much as living species can be classified by their evolutionary lineage.
12. Association for Pathology Informatics Information Resources, 2006. http://www.pathologyinformatics.org/informatics_r.htm. Comment. This Web site contains the latest versions of *The Developmental Lineage Classification and Taxonomy of Neoplasms.*
13. Berman, J. J. "Resources for comparing the speed and performance of medical autocoders." *BMC Medical Informatics and Decision Making* 4:8, 2004. http://www.biomedcentral.com/1472-6947/4/8. Comment. The field of machine translation cannot proceed unless there is a way for developers and end-users to perform side-by-side comparisons between different translators. If each developer of autocoding software is reluctant to provide his or her software to competitors, there should at least be an available baseline autocoder that can be used by everyone to express the relative speed and performance of their autocoder. This paper provides a minimalist autocoder (about 37 lines of Perl code) that can be used in benchmarking trials.
14. Grivell, L. "Mining the bibliome: searching for a needle in a haystack?" *EMBO Reports* 3:200–203, 2002. Comment. Grivell describes a technique for creating a text signature composed of a combination of terms and concepts extracted from the text, and serv-

ing as a metric against which other documents can be compared to determine the similarity between different documents. Similar documents can be retrieved based on relationships between their document signatures.

15. Heja, G. and G. Surjan. "Using n-gram method in the decomposition of compound medical diagnoses." *Int J Med Inf* 70:229-236, 2003. Comment. Some words appear frequently in combination with other words. The occurrence frequencies of two-word, three-word, and *n*-word phrases can be used to make inferences about the intended meaning of text.
16. Berman, J. J. "Doublet method for very fast autocoding." *BMC Medical Informatics and Decision Making* 4:16, 2004. http://www.biomedcentral.com/1472-6947/4/16. Comment. The doublet method is a lexical parsing algorithm that permits rapid autocoding with any nomenclature.
17. Berman, J. J. "Automatic extraction of candidate nomenclature terms using the doublet method." *BMC Medical Informatics and Decision Making* 5:35, 2005. Comment. This paper describes a quick computer program that will scan a corpus of text (of any size) and extract potential new terms for any given nomenclature.
18. Berman, J. J. "A Tool for Sharing Annotated Research Data: The 'Category 0' UMLS (Unified Medical Language System) Vocabularies" *BMC Medical Informantics and Decision Making* 3:6, 2003. http://www.biomedcentral.com/1472-6947/3/6. Comment. The largest curated listing of biomedical terms is the National Library of Medicine's Unified Medical Language System (UMLS). Many of the vocabularies contained in the UMLS carry restrictions on their use, making it impossible to share or distribute UMLS-annotated research data. However, a subset of the UMLS vocabularies, designated Category 0 by UMLS, can be used to annotate and share datasets without violating the UMLS License Agreement. This paper describes how to extract Category 0 vocabularies from UMLS.
19. Berman, J. J. "Confidentiality for Medical Data Miners." *Artificial Intelligence in Medicine* 26:25–36, 2002. Comment. This paper describes some of the innovative computational remedies that will permit researchers to conduct research and share their data without risk to patient or institution.
20. Berman, J. J. "Nomenclature-based data retrieval without prior annotation: facilitating biomedical data integration with fast doublet matching." In *Silico Biology* 5:0029, 2005. http://www.bioinfo.de/isb/2005/05/0029/. Comment. This article provides an algorithm and sample implementation that permits users to conduct concept retrievals (for all the terms in a nomenclature that are synonyms for a common concept) on-the-fly, using virtually any nomenclature.

21. SNOMED Clinical Terms. http://www.connectingforhealth.nhs.uk/technical/standards/snomed/. Comment. From the Web site: "The National Programme for Information Technology depends on having a common language for gathering and sharing medical knowledge. SNOMED CT will be the language of the NHS Care Records Service and will cut down the potential for differing interpretation of information and the possibility of errors resulting from traditional paper records. If clinical information is to be transferred and exchanged electronically, a standard clinical terminology is a necessary component of clinical systems. There would be problems in exchange of information for clinical or managerial purposes if several vocabularies and terms for the same topic were used within the NHS. SNOMED CT is, therefore, maintained and updated centrally. There will, however, be opportunities to submit requests for terms to be amended or introduced at a 'submission request' area on the NHS Terminology Service website."
22. National Library of Medicine. FAQs: Inclusion of SNOMED-CT in the UMLS. U.S. National Institutes of Health, 2003. http://www.nlm.nih.gov/research/umls/Snomed/snomed_faq.html. Comment. This FAQ explains encumbrances that apply to U.S. and non-U.S. users of SNOMED terms included in the UMLS Metathesaurus.
23. National Library of Medicine. SNOMED license agreement. U.S. National Institutes of Health, 2003. http://www.nlm.nih.gov/research/umls/Snomed/snomed_license.html. Comment. This is the agreement between the U.S. National Library of Medicine and the College of American Pathologists for the right to include SNOMED within the UMLS.
24. National Library of Medicine. SNOMED Clinical Terms (SNOMED-CT). U.S. National Institutes of Health, 2004. http://www.nlm.nih.gov/research/umls/Snomed/snomed_main.html. Comment. This is the only statement that I have found, issued by the National Library of Medicine, that plainly states that SNOMED is, in fact, a standard, "for use in U.S. Federal Government systems for the electronic exchange of clinical health information."
25. Schneier, B. *Applied Cryptography: Protocols, Algorithms and Source Code in C.* New York, Wiley. 1994. Comment. Bruce Schneier is one of the most influential and creative thinkers in the field of cryptography and computer security. This book provides an excellent discussion of the mathematical and technical aspects of the field.
26. Schneier, B. *Secrets and Lies. Digital security in a networked world.* Indianapolis, Wiley, 2000. Comment. This book delves into the social and human aspects of computer security. This is a very good companion to the same author's earlier book, *Applied Cryptography.*
27. Faldum, A, and K. Pommerening. "An optimal code for patient identifiers." *Comput Methods Programs Biomed* 79:81–88, 2005. Comment.

Without collisions, the authors propose a hash for distributing every patient of a billion patients using eight characters, two of which can be used for error detection/correction.

28. Rivest, R. "Request for Comments" 1321, The MD5 Message-Digest Algorithm. http://rfc.net/rfc1321.html. Comment. Also found at: http://theory.lcs.mit.edu/~rivest/Rivest-MD5.txt. Message-digest algorithms, also known as fingerprint algorithms or one-way hash algorithms, take a binary object, such as a file, and create a short, seemingly random string of characters (message digest) with the following properties: the same file will always produce the same message digest; if the file is changed in any way, the algorithm will produce a different message-digest; and it is impossible to compute the original file from any manipulation of the message digest. MD-5 is an open algorithm with easily obtained software implementations. It is used to authenticate binary objects.
29. "Secure Hash Algorithm (SHA-1)," *National Institute of Standards and Technology*, NIST FIPS PUB 180-1, "Secure Hash Standard," U.S. Department of Commerce, April 1995. http://www.itl.nist.gov/fipspubs/fip180-1.htm. Comment. The Secure Hash Algorithm is a message digest algorithm approved by the U.S. Department of Commerce.
30. Berman, J. J. "Threshold protocol for the exchange of confidential medical data." *BMC Medical Research Methodology* 2:12, 2002. http://www.biomedcentral.com/1471-2288/2/12. Comment. Threshold cryptographic protocols divide messages into multiple pieces, with no single piece containing information that can reconstruct the original message. This paper describes a novel threshold protocol that can be used to search, annotate, or transform confidential data without breaching patient confidentiality. The threshold algorithm produces two files (threshold pieces). In typical usage, the data owner holds Piece 2, and Piece 1 is freely distributed. Piece 1 can be annotated and returned to the owner of the original data to enhance the complete dataset. Collections of Piece 1 files can be merged and distributed without identifying patient records. Variations of the threshold protocol are described. The author's Perl implementation is freely available.
31. Department of Health and Human Services. "45 CFR (Code of Federal Regulations), Parts 160 through 164. Standards for Privacy of Individually Identifiable Health Information (Final Rule)." *Federal Register*, Volume 65, Number 250, pp. 82461–82510, December 28, 2000. http://aspe.hhs.gov/admnsimp/final/PvcPre01.htm. Comment. This is the much-feared HIPAA rule for protecting patient privacy.
32. Department of Health and Human Services. "45 CFR (Code of Federal Regulations), 46. Protection of Human Subjects (Common Rule)." *Federal Register*, Volume 56, pp. 28003–28032, June 18, 1991. http://www.hhs.gov/ohrp/humansubjects/guidance/45cfr46.htm.

Comment. Although HIPAA (Health Insurance Portability and Accountability Act) privacy regulations seem to get all the attention, the Common Rule sets the basic principles for protecting patients from research risks, mandating the activities of Institutional Review Boards, and using human tissues in support of medical research. It is essential reading for anyone involved in human subject research.

33. Malin, B. and L. Sweeney. "How (not) to protect genomic data privacy in a distributed network: using trail re-identification to evaluate and design anonymity protection systems." *J Biomed Inform* 37:179–192, 2004. Comment. Latanya Sweeny was one of the first people to point out the difficulties in achieving de-identification of medical records. She has not published many papers, but each publication is worth reading.
34. Sweeney, L. "Guaranteeing anonymity when sharing medical data, the Datafly system." *Proc American Medical Informatics Association* pp. 51–55, 1997. Comment. Latanya Sweeney was one of the first researchers to show that personal data extracted from different databases can be connected to identify patients whose names were removed from clinical research datasets.
35. Sweeney, L. "Three computational systems for disclosing medical data in the year 1999." *Medinfo.* 9(Pt 2), pp. 1124–1129, 1998. Comment. This is another contribution from Latanya Sweeney, who has pioneered computational methods for safely disclosing medical data without breaching patient confidentiality.
36. Berman, J. J. "Racing to share pathology data." *Am J Clin Pathol* 121:169–171, 2004. Comment. Patents are being sought on some of the basic methods for sharing biomedical data.
37. Berman, J. J. "Concept-Match Medical Data Scrubbing: How pathology datasets can be used in research." *Arch Pathol Lab Med Arch Pathol Lab Med*, 127:680–686, 2003. Comment. Medical data scrubbing involves the removal of words from free text that can be used to identify persons or that contain information that is incriminating or otherwise private or information that does not meet the "minimum necessary" directive under HIPAA. This article describes a general algorithm that scrubs pathology free text quickly and without error. Essentially, all words in text are blocked out unless they are high-frequency words (such as, of, it, the, when) and words that match terms within a medical nomenclature. The scrubbed output preserves the sense of the original sentences.
38. Cho, M. K., S. Illangasekare, M. A. Weaver, D. G. B. Leonard, and J. F. Merz. "Effect of patents and licenses on the provision of clinical genetic testing services." *J Mol Diag* 5:3–8, 2003. Comment. When a patent encumbers a clinical test, clinical laboratory directors might be reluctant to provide the test in their laboratories. In a telephone survey, 53 percent of clinical laboratory directors said they decided against developing a new clinical genetic test because of a patent or license issue.

39. Merz, J. F., A. G. Kriss, D. G. Leonard, and M. K. Cho. "Diagnostic testing fails the test." *Nature* 415:577–579, 2002. Comment. Once a patent is issued for a diagnostic test, laboratories often choose not to offer the test rather than pay royalties to the patent holder.
40. Wong, C. *Web Client Programming with Perl.* O'Reilly, Sebastopol, CA, 1997. Comment. This somewhat-dated book provides a clear description of popular Perl methods for extracting data from the Internet.
41. Berman, J. J. and K. Bhatia. "Biomedical Data Integration: Using XML to Link Clinical and Research Datasets." *Expert Reviews in Molecular Diagnostics* 5:329–336, 2005. Comment. Review article extolling the virtues of XML for integrating heterogeneous biomedical data.
42. Berman, J. J. "Pathology Data Integration with eXtensible Markup Language." *Human Pathology* 36:139–145, 2005. Comment. Autopsy and surgical pathology reports have particular importance for biomedical professionals because they contain a retrievable diagnosis based on observations made on retrievable archived tissue blocks. This article describes how pathology data can be integrated with biologic datasets, using XML.
43. ISO/IEC 11179. http://en.wikipedia.org/wiki/ISO-11179. Comment. ISO 11179 is a "Specification and Standardization of Data Elements." It has been considered an unnecessary metadata standard and is often ignored in the XML community. However, ISO11179 is a crucial, fundamental part of any informatics effort. It provides a formal mechanism for creating unique global identifiers for standard data elements and provides uniform guidance for identifying, developing, and describing data elements. If all metadata had formal descriptions conforming to ISO11179, standardized data could be exchanged between different organizations and the enormous redundancy of in-use data elements would be reduced or eliminated.
44. White, C., L. Quin, and L. Burman. *Mastering XML: Premium Edition.* Sybex, San Francisco, 2001. Comment. This book, and its updated editions, is an excellent review of XML principles and tools. It has an exceptionally useful section on Internet standards, pp. 224–231.
45. Brazma, A., P. Hingamp, J. Quackenbush, G. Sherlock, P. Spellman, C. Stoeckert, J. Aach, W. Ansorge, C. A. Ball, H. C. Causton, et al. "Minimum information about a microarray experiment (MIAME) - toward standards for microarray data." *Nat Genet* 29:365–371, 2001. Comment. Defines the content and structure of a minimal set of information needed to describe a microarray dataset.
46. Spellman, P. T., M. Miller, J. Stewart, C. Troup, U. Sarkans, S. Chervitz, D. Bernhart, G. Sherlock, C. Ball, et al. "Design and implementation of microarray gene expression markup language (MAGE-ML)." *Genome Biol* 3:0046, 2002. Comment. This paper describes the gene expression microarray mark-up language and is one of the

earliest and best examples of XML schema specifications for emerging biomedical technologies.

47. Berman, J. J., M. E. Edgerton, and B. Friedman. "The Tissue Microarray Data Exchange Specification: A Community-based, Open Source Tool for Sharing Tissue Microarray Data." *BMC Medical Informatics and Decision Making* 3:5, 2003. http://www.biomedcentral.com/1472-6947/3/5. Comment. Tissue Microarrays (TMAs) allow researchers to examine hundreds of small tissue samples on a single glass slide. The information held in a single TMA slide might easily involve gigabytes of data. To benefit from TMA technology, the scientific community needs an open source TMA data-exchange specification that will convey all of the data in a TMA experiment in a format that is understandable to both humans and computers. A data-exchange specification for TMAs allows researchers to submit their data to journals and to public data repositories and to share or merge data from different laboratories.
48. M. A. Harris, J. Clark, A. Ireland, et al. "Gene Ontology Consortium. The Gene Ontology (GO) database and informatics resource." *Nucl Acids Res* 32:D258–D261, 2004. http://nar.oxfordjournals.org/cgi/content/full/32/suppl_1/D258. Comment. GO can be thought of as three classifications merged into an ontology: molecular functions, biological processes, and cellular components. Molecular function describes molecule-level activities such as a specific catalytic or binding action. An example is "kinase activity." Biological process describes coordinated cellular activities (for example, respiration, cell death) that involve macromolecules. Cellular component describes subcellular locations of molecular activities and structures.
49. Futreal, P. A., L. Coin, M. Marshall, T. Down, T. Hubbard, R. Wooster, N. Rahman, and M. R. Stratton. "A census of human cancer genes." *Nature Reviews Cancer* 4:177–183, 2004. Comment. The authors have carefully collected a comprehensive list of human oncogenes.
50. The Wellcome Trust Sanger Institute Census of Cancer Genes. http://www.sanger.ac.uk/genetics/CGP/Census/. Comment. This site reposits the latest version of the oncogene census of human cancers.
51. Ahmed, K., D. Ayers, M. Birbeck, J. Cousins, D. Dodds, J. Lubell, M. Nic, D. Rivers-Moore A. Watt, R. Worden, and A. Wrightson. *Professional XML Meta Data.* Wrox Press Ltd., Birmingham, 2001. Comment. Metadata is the building block of XML, the semantic Web, ontologies, software agents, and just about every projected Web technology. Every informatician should have a deep understanding of metadata, and this book is an excellent resource.
52. *OWL Web Ontology Language Reference.* W3C Recommendation, 10 February 2004. http://www.w3.org/TR/owl-ref/. Comment. This is the official W3C reference for OWL, a formal specification for a set of RDF triples that provide logical inferencing structures for ontologies.

53. POSIX—Perl interface to IEEE Std 1003.1. www.fnal.gov/docs/products/perl/pod.new/5.00503/i686-linux/POSIX.html. Comment. POSIX is a suite of fundamental operations, many mathematical, that any programming language might include. The same POSIX routine should perform identically in any conforming programming language, using the same name for the routine.
54. Orwant, J., J. Hietaniemi, and J. Macdonald. *Mastering Algorithms with Perl.* O'Reilly, Sebastopol, California, 1999. Comment. This is an excellent book that clearly describes commonly used algorithms. Working Perl scripts are provided.
55. Number crunching capabilities for Perl. http//pdl.perl.org. Comment. The PDL (Perl data language) Web site.
56. Swets, J. A., R. M. Dawes, and J. Monahan. "Better decisions through science." *Scientific American*, 82–87, October, 2000. Comment. Clear and concise explanation of Receiver Operator Characteristics.
57. Kestler, H. A. "ROC with confidence - a Perl program for receiver operator characteristic curves." *Comput Methods Programs Biomed* 64:133–136, 2001. Comment. ROC curves are used in biomedicine to evaluate predictive tests.
58. Berman, J. J. and G. W. Moore. "The role of cell death in the growth of preneoplastic lesions: a Monte Carlo simulation model." *Cell Proliferation* 25:549–557, 1992. Comment. Regression of precancerous lesions is common. This paper examines the hypothesis that early lesions operate under the identical growth kinetics of 'late' lesions (neoplasms), but that kinetic features favoring continuous growth in established lesions tend to favor extinction of lesions composed of small numbers of cells. Growth simulations of early lesions were produced using the Monte Carlo method. The model demonstrates that small increments in the intrinsic cell loss probability in even the earliest progenitors of malignancy can strongly influence the subsequent development of neoplasia from initiated foci.
59. Berman, J. J. and G. W. Moore. "Spontaneous regression of residual tumor burden: prediction by Monte Carlo Simulation." *Anal Cellul Pathol* 4:359–368, 1992. Comment. This manuscript finds a plausible explanation for the clinically observed failure of tumors to recur in instances where tumor burden remains following cancer therapy. The paper also shows that the Monte Carlo method can simulate biologic events in populations when the fate of each member of a population can be modeled probabilistically.
60. Day, R. S., S. E. Shacknew, and W. P. Peters. "The analysis of relapse-free survival curves: implications for evaluating intensive systemic adjuvant treatment regimens for breast cancer." *Br J Cancer* 92:47-54, 2005. Comment: The authors develop a natural history model for the time to local tumor recurrence of breast cancer.

61. Berman, J. J. "Predicting tumor marker outcomes with Monte Carlo simulations." APIII, Pittsburgh, October 7, 2003. apiii.upmc.edu/abstracts/posterarchive/2003/abs_include.cfm?file=berman1.html. Comment. Computer models can be used to anticipate and avoid design problems in expensive clinical trials.
62. Stallman, R. Why "Free Software" is better than "Open Source." http://www.gnu.org/philosophy/free-software-for-freedom.html. Comment. In practice, there is very little difference between the free software movement and the open source initiative. The essay by Richard Stallman, guru of the free software movement, stresses philosophic differences between these two related projects.
63. Comprehensive Perl Archive Network. http://www.cpan.org/. Comment. This site archives over 9,000 Perl modules that can be freely downloaded. These modules greatly extend the utility of Perl.
64. de Hoon, M., S. Imoto, and S. Miyano. The C Clustering Library for cDNA microarray data. University of Tokyo, 2005. http://bonsai.ims.u-tokyo.ac.jp/~mdehoon/software/cluster/cluster.pdf. Comment. This open access suite of clustering algorithms can be extended to many types of multiplexed datasets. Instructions are included for implementing the algorithms from a Perl script.
65. St. Laurent, S., J. Johnston, and E. Dumbill. *Programming Web Services with XML-RPC.* O'Reilly, Sebastopol, CA, 2001. Comment. This is a nice book on remote procedure calling using XML syntax.
66. Conway, D. *Object Oriented Perl.* Manning, Greenwich, 2000. Comment. Object-oriented Perl programming is difficult to learn, but Damian Conway's book is easy to read. This book is worth reading and rereading.
67. Berman, J. J. "Specialized cancer nomenclatures." Submitted, *Cancer Informatics*, Dec. 27, 2005. Comment. General nomenclatures might lack terms and concepts found in nomenclatures that focus on a narrow subdomain included in the general nomenclature. This means that for certain types of studies, particularly those that need to retrieve rare entities or uncommon variants of common entities, medical codes using standard nomenclatures need to be supplemented with codes from specialized nomenclatures.
68. Wang, X., R. Gorlitsky, and J. S. Almeida. "From XML to RDF: how semantic Web technologies will change the design of 'omic' standards." *Nature Biotechnology* 23:1099–1103, 2005. http://www.nature.com/nbt/journal/v23/n9/full/nbt1139.html. Comment. The authors suggest that XML schemas do a poor job of representing data standards. The authors prefer the use of RDF, which provides a simple data model in which assertions (in the form of triples) can be easily exchanged between heterogeneous data documents.
69. Volume 34, *Database Issue. Nucleic acids Research.* January 1, 2006. http://nar.oxfordjournals.org/content/vol34/suppl_1/index.dtl.

Comment. The yearly biology database issue lists over 800 databases. The descriptions of the different databases are available as open access articles.

70. Shibuya, K., C. D. Mathers, C. Boschi-Pinto, A. D. Lopez, and C. J. L. Murray. "Global and regional estimates of cancer mortality and incidence by site: II. Results for the global burden of disease, 2000." *BMC Cancer* 2002, 2:37. Comment. The types of tumors that occur in developing nations differs substantially from the tumors that occur in many Western countries. This paper is an important resource for worldwide cancer incidence, listed by country and tumor site.
71. Frey, C. M., M. M. McMillen, C. D. Cowan, J. W. Horm, and L. G. Kessler: "Representativeness of the surveillance, epidemiology, and end results program data: recent trends in cancer mortality rate." *JNCI* 84:872, 1992. Comment. The Surveillance, Epidemiology, and End Results (SEER) project collects cancer-related incidence and mortality data from residents in geographically defined populations, representing about 10 percent of the U.S. population.
72. National Cancer Institute Factsheet. "Bone Cancer: Questions and Answers." http://www.nci.nih.gov/cancertopics/factsheet/Sites-Types/bone. Comment. In 2000, there were about 2,500 new cases of bone sarcoma in the United States.
73. Gentleman, R. C., D. M. Bates, B. Bolstad, V. J. Carey, M. Dettling, S. Dudoit, B. Ellis, L. Gautier, Y. Ge, J. Gentry, et al. "Bioconductor: open software development for computational biology and bioinformatics." *Genome Biol* 5:R8, September 15, 2004. http://genomebiology.com/2004/5/10/R80. Comment. The Bioconductor project is a collection of open source software for computational biologists. It draws heavily on the statistics programming language, R.
74. Dublin Core Metadata Initiative. http://dublincore.org/. Comment. The Dublin Core (see Glossary) is a set of basic metadata that describe XML documents. The Dublin Core were developed by a forward-seeing group of library scientists who understood that every XML document needs to include self-describing metadata that will allow the document to be indexed and appropriately retrieved.
75. Dublin Core Metadata Element Set, Version 1.1: Reference Description. http://dublincore.org/documents/1999/07/02/dces/. Comment. This is the URL for the Dublin Core elements.
76. What is copyleft? http://www.gnu.org/copyleft/. Comment. The GNU organization publishes two licenses used for software produced by GNU and by anyone who would like to distribute their software under the terms of the GNU license. The licenses are referred to as copyleft licenses, because the licenses primarily serve the software users, rather than the software creators. One of the GNU licenses, the General Public License, covers most software applications. The GNU Lesser General Public License, formerly known as the GNU Library

General Public License, is intended for use with software libraries or unified collections of files constituting a complex application, language, or other body of work.

77. Guidance for Industry, FDA Reviewers, and Compliance on Off-The-Shelf Software Use in Medical Devices. September 9, 1999. http://www.fda.gov/cdrh/ode/guidance/585.pdf. Comment. Off-the-shelf software may be used in new medical devices as long as the device manufacturer follows FDA guidance.
78. Open Source Definition. http://www.opensource.org/docs/definition_plain.php. Comment. This definition is more akin to an open source "Bill of Rights." The first draft of the current definition is attributed to Bruce Perens, one of the principle players in the open source movement.
79. Open Source Initiative License Index. http://www.opensource.org/licenses/. Comment. The Open Source Initiative has an approval process for open source licenses. Software distributed under an approved license can include a declaration that the software is "OSI Certified Open Source Software." The GNU copyleft licenses have been certified as open source software licenses.

Resources

OPEN SOURCE BIOPERL, BIOJAVA, BIOPYTHON, BIORUBY

Beginning with the venerable BioPerl organization, the following language-specific bioinformatics efforts each provide open source environments supporting a range of bioinformatics tasks:

http://www.bioperl.org/

http://www.biojava.org/

http://www.biopython.org/

http://www.bioruby.org/

OPEN SOURCE COMPRESSION AND ARCHIVING UTILITIES (GZIP, GUNZIP, TAR, 7-ZIP, BUNZIP)

Gzip (short for GNU zip) is a compression utility designed as a surrogate for "compress," another popular compression utility that uses a patented algorithm. It has been adopted by the GNU project. Gzipped files are very popular among Linux users and are widely prevalent on Internet download sites.

Gunzip is the decompression utility that operates on gzipped files.

Gzip and gunzip are available for no-cost download from:

www.gzip.org

TAR (originally Tape ARchiver) has outlasted its original purpose. TAR works on multiple platforms to archive multiple files into a single file for easy storage and transport. TAR will de-archive the files on command.

Many people will compress the TAR archive using gzip, and it is common to find files with a double suffix:

filename.tar.gz

When you consecutively decompress with gunzip and de-archive with TAR, you typically get a library of files distributed in a pre-named directory (and subdirectories). TAR can be downloaded from:

http://www.gnu.org/software/tar/tar.html

7-zip is a compression/decompression and archiving/de-archiving utility. It will decompress files created with commercial software.

A download of this wonderful utility, distributed under a GNU LGPL license, is available at:

http://www.7-zip.org/

Bunzip is a free implementation of the clever Burrows-Wheeler transform. For certain types of data, and with ample RAM, Bunzip can provide higher compression than many other algorithms.

Most users of the algorithms use the newer `bunzip2` version:

http://www.bzip.org/

R OPEN SOURCE STATISTICAL PROGRAMMING LANGUAGE AND BIOCONDUCTOR

R is an open source programming language for statistics. A free, easy-to-install version for non-Linux users can be found at:

http://www.stats.bris.ac.uk/R/bin/windows/base/

A general R FAQ is available at:

http://cran.r-project.org/doc/FAQ/R-FAQ.html

R is an example of a successful, sophisticated software project that was completed over several years by a group of dedicated experts scattered across the globe. The effort had virtually no funding.

Bioconductor is an open source software environment for computational biology and bioinformatics. It uses R extensively. Robert Gentleman, a contributor to R, is primary author of an article on Bioconductor (73).

CYGWIN, OPEN SOURCE UNIX/LINUX EMULATOR

Cygwin emulates many of the common UNIX/Linux commands, for Windows users.

Cygwin can be downloaded from:

http://www.cygwin.com/

Download `setup.exe`.

When the dialog box displays available components, expand the "development tools" box and check off all the `gcc` components. This will provide you with GNU's versions of C and C++.

GNUPG—OPEN SOURCE ENCRYPTION TOOL

GnuPG, the GNU Privacy Guard, is a free replacement for PGP.

PGP (Pretty Good Privacy) has a colorful history. Phil Zimmermann labored to develop PGP as a freeware encryption program, and was accused by the U.S. government of being an exporter of munitions. PGP and other encryption software were considered munitions in the 1990s. Though this seems strange, it should be remembered that, for decades, mathematicians employed by the National Security Agency were the near-exclusive developers of encryption algorithms. The appearance and proliferation of innovative public encryption software shocked U.S. intelligence agencies. The charges against Zimmermann were eventually dropped. Today, PGP is a commercial software company.

Those who prefer free and open source cryptography software might consider acquiring GNU's Privacy Guard. Because GnuPG does not use the patented IDEA algorithm, it can be used without any restrictions.

GnuPG is a RFC2440 (OpenPGP) compliant application:

http://www.faqs.org/rfcs/rfc2440.html

GnuPG is available for several popular operating systems and can be downloaded from:

http://www.gnupg.org/download/

For those seeking simplicity, GnuPG can be used to produce a simple symmetric encryption of a file, using the -c command from the DOS command line, as shown:

```
C:\gpgdos>gpg -c outline.txt
```

The file `outline.txt` is a sample for encryption/decryption, and `gpgdos` is simply a created subdirectory containing the GPG files.

You will be prompted for a password. The encrypted output file produced by GnuPG is: `outline.txt.gpg`.

The encrypted file can be decrypted with the -d decryption command and redirected to the output file named `out.out`:

```
C:\gpgdos>gpg -d outline.txt.gpg >c:\gpgdos\out.out
```

You will be prompted again for the same password used as the key for the encryption step.

WGET WEB SITE MIRRORING SOFTWARE

Wget is a freely available command-line utility that can be used to download Web pages or entire Web sites.

Wget is the non-Linux version of the Linux utility "get" and can be downloaded from: *ftp://sunsite.dk/projects/wget/windows/*

A tutorial on `wget` is located at:

http://www.gnu.org/software/wget/wget.html

CWM—A CLOSED WORLD MACHINE FOR RDF (IN PYTHON)

CWM (Closed World Machine) is a command-line Python program that will interconvert RDF and notation 3 (n3) files. It is available at:

http://www.w3.org/2000/10/swap/doc/CwmInstall

CWM will convert an RDF file to n3, as shown in this command-line example:

```
C:\cwm-1.0.0>cwm —n3 sample.rdf
```

CWM also will logically merge multiple n3 files, as shown in this command-line example:

```
c:\cwm-1.0.0>cwm one.n3 two.n3 three.n3
```

DATABASE ISSUE OF NUCLEIC ACIDS RESEARCH

The most current database issue of Nucleic Acids Research can be found at:

http://nar.oxfordjournals.org/content/vol35/suppl_1/index.dtl

The 2007 issue describes 174 biological databases.

UMLS METATHESAURUS

By far, the largest medical nomenclature is the U.S. National Library of Medicine's Unified Medical Language System Metathesaurus (UMLS Metathesaurus), which contains several million terms grouped under about a million concepts. Over 100 individual thesauruses (see Glossary) comprise the UMLS, many of which are contributed by efforts funded by the U.S. government. The UMLS metathesaurus can be acquired at no cost from the National Library of Medicine.

To get UMLS files:

- Get a free UMLS license. Fill out your application at: http://www.nlm.nih.gov/research/umls/license.html
- This will eventually result in your having a license number (which you might never need), and a login ID and a password (both of which you will definitely need whenever you want to download UMLS metathesaurus files). Once you have a login ID and password, go to: *http://umlsks.nlm.nih.gov/kss/servlet/Turbine/template/admin,user,KSS_login.vm*
- Enter your login ID and password.This takes you to the UMLSKS server. On the left side of the Web page, go to downloads and click on UMLS Knowledge Sources. Download the latest metathesaurus file collections. In 2006, the metathesaurus was contained in three files with a `.nlm` extension.

The `.nlm` extension is a zipped archive. Rename the files with `.zip` extensions. Use a zip utility, such as the freely available 7-zip to decompress the latest metathesaurus files.

- You will see that the zip-archived files are individual gzipped (`.gz`) files. Two of the most useful metathesaurus files are the MRCONSO file (compressed as `MRCONSO.RRF.GZ` and referred to as the MRCON file in pre-2006 versions of the metathesaurus), containing all the UMLS terms, and the MRSTY file (compressed as `MRSTY.RRF.GZ`), which contains the relationships (concept class) for the terms contained in the MRCONSO file.

The following is an example of the first record in the MRCONSO file.

C0000005|ENG|P|L0000005|PF|S0007492|Y|A7755565||M0019694|D012711|MSH|PEN|D012711|(131)I-Macroaggregated Albumin|0|N||

In this record, the UMLS unique concept code is "C0000005." The term is "(131)I-Macroaggregated Albumin."

The following are a few records from the MRSTYR file:

C0000741|T023|A1.2.3.1|Body Part, Organ, or Organ Component|AT17732098|

C0000742|T005|A1.1.3|Virus|AT17595612||

C0000744|T047|B2.2.1.2.1|Disease or Syndrome|AT17683825||

C0000754|T019|A1.2.2.1|Congenital Abnormality|AT08212029||

TAXONOMY

One of the best examples of a large taxonomy (see Glossary) is `taxonomy.dat`, which lists the organisms sampled by bioinformatics databases.

So thorough is `taxonomy.dat` that it not only lists all known variations of an organism's name, it also lists commonly used misspellings of an organism (List R.1).

LIST R.1. A RECORD IN TAXONOMY

```
ID : 50
PARENT ID : 49
RANK: genus
GC ID : 11
SCIENTIFIC NAME : Chondromyces
SYNONYM : Polycephalum
SYNONYM : Myxobotrys
SYNONYM : Chondromyces Berkeley and Curtis 1874
SYNONYM : "Polycephalum" Kalchbrenner and Cooke 1880
SYNONYM : "Myxobotrys" Zukal 1896
MISSPELLING : Chrondromyces
```

The `taxonomy.dat` file is available for public download through anonymous ftp:

ftp://ftp.ebi.ac.uk/pub/databases/taxonomy/

Information about the `taxonomy.dat` file is found at:

http://www.ebi.ac.uk/msd-srv/docs/dbdoc/ref_taxonomy.html

LOCUSLINK AND ITS SUCCESSOR, ENTREZ GENE

The story of LocusLink illustrates some of the legacy-related difficulties encountered in the biomedicine community.

LocusLink was a database sponsored by the U.S. National Library of Medicine. It provided curated sequences and a range of descriptive information about genetic loci. Rich linkage to OMIM and to literature citations made it one of the few truly biomedical genomic databases (that is, a database that associates biological data with clinical data).

LocusLink data was assembled in a large, curated data file, `LL_TMPL`, which is available by anonymous ftp from:

ftp.ncbi.nih.gov, at subdirectory /refseq/LocusLink/ARCH

An abridged example of the first LocusLink record is:

```
>1
LOCUSID: 1
LOCUS_CONFIRMED: yes
LOCUS_TYPE: gene with protein product, function known or
   inferred
ORGANISM: Homo sapiens

.

.
OFFICIAL_SYMBOL: A1BG
OFFICIAL_GENE_NAME: alpha-1-B glycoprotein
ALIAS_SYMBOL: A1B
ALIAS_SYMBOL: ABG
ALIAS_SYMBOL: GAB
ALIAS_SYMBOL: HYST2477
ALIAS_SYMBOL: DKFZp686F0970
PREFERRED_PRODUCT: alpha 1B-glycoprotein
SUMMARY: The protein encoded by this gene is a
   plasma glycoprotein of unknown function. The protein
   shows sequence similarity to the variable regions of
   some immunoglobulin supergene family member proteins.
CHR: 19

 .

 .
ALIAS_PROT: alpha-1B-glycoprotein
LINK: http://www.ncbi.nlm.nih.gov/UniGene/
   clust.cgi?ORG=Hs&CID=529161
UNIGENE: Hs.529161
OMIM: 138670
MAP: 19q13.4|RefSeq|C|
MAPLINK: default_human_gene|A1BG
LINK: http://www.ncbi.nlm.nih.gov/SNP/
   snp_ref.cgi?locusId=1
```

```
LINK: http://www.ncbi.nlm.nih.gov/HomoloGene/
   homolquery.cgi?TEXT=1[loc]&TAXID=9606
LINK: http://www.gdb.org/gdb-bin/genera/accno?GDB:119638
LINK: http://www.ensembl.org/Homo_sapiens/
   contigview?geneid=AK055885
LINK:  http://genome.ucsc.edu/cgi-bin/
   hgTracks?org=human&position=AK055885
PMID: 15461460,15221005,14702039,12477932,8889549,
   3610142,3458201,2591067
GO: biological process|biological process
   unknown|ND|GO:0000004|GOA/IPI|na
GO: cellular component|extracellular
   region|IDA|GO:0005576|GOA/IPI|3458201
GO: molecular function|molecular function
   unknown|ND|GO:0005554|GOA/IPI|3458201
```

By linking sequences to standard nomenclatures, to function, to Gene Ontology (GO) entries, to OMIM, and to PubMed citations, LL_TMPL was one of the richest public sources of integrated biomedical information in existence.

The latest LL_TMPL file is 245,045,067 bytes in length.

In 2005, the NCBI transitioned LocusLink into Entrez Gene. The information that was contained in LL_TMPL is now distributed in a list of downloadable Entrez data files, also available through anonymous ftp:

```
ftp.ncbi.nih.gov, at subdirectory /gene/DATA
Feb 14, 2006 gene2accession.gz 63681288
Feb 14, 2006 gene2go.gz          5520028
Feb 14, 2006 gene2pubmed.gz      7565480
Feb 14, 2006 gene2refseq.gz     36697899
Feb 14, 2006 gene2sts            3611571
Feb 14, 2006 gene2unigene        3439332
Feb 14, 2006 gene_history.gz     1500619
Feb 14, 2006 gene_info.gz       27857831
Feb 14, 2006 mim2gene             256715
```

When a database transitions, new software is required to retrieve and integrate data from the new data sources. The skills needed to provide the new software are really simple: knowledge of the kinds of data available in the new dataset; knowledge of the manner in which the new dataset is organized and structured; and a few simple methods for parsing through the new datasets to conduct data searches or data transformations (that is, reorganizing or restructuring the data in a preferred format). So long as datasets undergo structural modifications and enhancements, biomedical professionals will make adjustments.

The most relevant file is `gene_info.gz`, which contains free-text descriptions of genes, including relations to tumors.

SOURCEFORGE

SourceForge describes itself as: "The world's largest development and download repository of Open Source code and applications. Providing free services to Open Source developers."

SourceForge is a remarkable communitarian artifice that hosts thousands of active, cooperative, open source efforts and provides public downloads of developed software. All software developed through SourceForge must be released under an open source license. The SourceForge Web site has full documentation for development teams and can be found at:

http://sourceforge.net

The SourceForge group-development process is managed with CVS, an open source tool that permits revision management on multiauthored software. SourceForge.net provides a File Release System to distribute releases to end-users.

CVS, CONCURRENT VERSIONS SYSTEM

CVS is an open source tool that keeps track of changes in software or documents. With CVS, you can work with colleagues on a software project, with each developer contributing concurrent edits to the project. If a new version of the software is problematic, the group can use CVS to revert to an earlier version of the software.

CVS is available for free download from:

http://www.nongnu.org/cvs/

CPAN—THE COMPREHENSIVE PERL ARCHIVE NETWORK

CPAN is the Comprehensive Perl Archive Network. It consists of thousands of Perl modules contributed by volunteer programmers. All of the Perl software at the CPAN site can be freely downloaded at:

http://www.cpan.org

OMIM—ONLINE MENDELIAN INHERITANCE IN MAN

OMIM is a listing of every known inherited condition in humans. Each condition has biologic and clinical descriptions in a detailed textual narrative that includes a listing of relevant citations. There are nearly 17,000 conditions described, and the OMIM file exceeds 100 megabytes. The OMIM file can be downloaded from the National Center for Bioinformatics anonymous ftp site: *ftp://ftp.ncbi.nih.gov and subdirectory: /repository/omim/*

Additional information on OMIM is available at:

http://www.ncbi.nlm.nih.gov/omim/

LOINC—LOGICAL OBSERVATIONS, IDENTIFIERS, NAMES, AND CODES

The intended purpose of LOINC is to provide a consistent method for describing clinical laboratory tests.

The LOINC file is about 13 megabytes in size and contains about 35,000 records. LOINC is freely downloadable from: *http://www.regenstrief.org/loinc*

LOINC is registered and copyrighted to the Regenstrief Institute, Inc. An abbreviated LOINC record (10445-5) is shown:

```
"10445-5"
"CD11C AG"
"ACNC"
"PT"
"TISS"
"ORD"
"IMMUNE STAIN"
"LEU-M5"
"CELLMARK"
"DL-M"
"20000830"
"Edited related names;changed component from LEU-M5 AG"
"NAM"
"CR4; AlphaX integrin chain; Axb2; leukocyte surface
   antigen p150,95; Leu-M5; LeuM5; Tissue; Antigens;
   Imun; Imune; Imm"
"CD11C Ag Tiss Ql ImStn"
```

HL7—HEALTH LEVEL 7

HL7 is a specification for transporting certain types of data that are typically collected by hospitals. Collections of HL7 data are called messages and are used to encapsulate information such as:

- Registration data: patient admissions, discharges, transfers
- Results/observations: laboratory tests, diagnoses, clinical observations, operative notes, large amounts of text
- Orders: from pharmacy, laboratory, or nurse
- Billing/charges

Virtually any kind of information can be packaged into an HL7 message. All HL7 messages have a similar structure.

The message begins with a header that contains a description of the message contents. A Patient Identification (PID) segment containing patient demographics typically follows the message header.

The PID is followed by observations segments (OBR), which are composed of individual observations (OBX) delimited by a vertical character, " | ".

OBX segments typically specify the test identifier, description, value and code source, unit of measure, and observed value status for a test.

SEER

The Surveillance, Epidemiology, and End Results (SEER) Program of the National Cancer Institute is an authoritative source of information on cancer incidence and survival in the United States. The SEER Public-Use Data include SEER incidence and population data associated by age, gender, race, year of diagnosis, and geographic areas (including SEER registry and county). Public-Use SEER data are available at:

http://seer.cancer.gov/publicdata/

The public can download 113 Megabytes of compressed SEER data (expands to 700 Megabytes). Users must sign a Public-Use Data Agreement before downloading SEER Public-Use files. The DATA Agreement is available at:

http://seer.cancer.gov/publicdata/access.html

MEDICAL SUBJECT HEADINGS—MESH

MESH (Medical Subject Headings) is a large, comprehensive, curated vocabulary created and maintained by the U.S. National Library of Medicine. It contains descriptors in a hierarchical structure (usually referred to as a MESH tree) that allows searching at all levels of granularity.

Information on MESH can be found at:

http://www.nlm.nih.gov/mesh/meshhome.html

The MESH download page can be found at:

http://www.nlm.nih.gov/mesh/filelist.html

The Jan 10, 2006 downloadable MESH flat file is named `D2006.bin` and is over 25 megabytes in length. It contains nearly 23,886 records, and each record has a similar form. An abbreviated record for "heparin" is shown:

```
*NEWRECORD
RECTYPE = D
MH = Heparin



.
.
.
MS = A highly acidic mucopolysaccharide formed of equal
parts of sulfated D-glucosamine and D-glucuronic acid
with sulfaminic bridges. The molecular weight ranges
from six to twenty thousand. Heparin occurs in and is
obtained from liver, lung, mast cells, etc., of
vertebrates. Its function is unknown, but it is used
to prevent blood clotting in vivo and vitro, in the
form of many different salts.
PM = /therapeutic use was HEPARIN, THERAPEUTIC 1965
HN = /therapeutic use was HEPARIN, THERAPEUTIC 1965
.
.
.

UI = D006493
```

GENE ONTOLOGY—GO

The GO Consortium Web site can be found at:

http://www.geneontology.org/

The GO Consortium distributes its products without a license, so long as users comply with its redistribution and citation policy, which can be found at:

http://www.geneontology.org/doc/GO.cite.html

The latest versions of GO can be downloaded from:

http://archive.godatabase.org/latest-termdb/

OBO (OPEN BIOLOGY ONTOLOGIES)

OBO is an open source collection of biological ontologies.

At the time of this writing, most of the ontologies were devoted to animals (other than human beings). The OBO Web site can be found at:

http://obo.sourceforge.net/browse.html

NEOPLASM CLASSIFICATION

The Developmental Lineage Classification and Taxonomy of Neoplasms contains over 145,000 names of neoplasms and has been described in several open access publications (9), (10), (11), (17). It is the world's largest source of names of human neoplasms. At the time of this writing, the most recent public version was available from the Association for Pathology Informatics Web site (12):

www.pathologyinformatics.org/informatics_r.htm

The Classification is an example of a comprehensive listing of all items in a circumscribed field of knowledge. Aside from its value to cancer researchers, it has value for software developers who are developing or testing autocoders, lexical parsers, term extractors, and a variety of data-mining projects. As an XML document, it can be easily parsed and transformed into any preferred format.

The neoplasm nomenclature is distributed in three gzipped files. `Neoclxml.gz` should be renamed `neocl.xml` when decompressed. `Neoself.gz`, upon decompression, becomes a flat file that lists each term's class ancestry.

The file `Neocl.txt` corresponds to the unclassified names of neoplasms contained in the nomenclature and can be found at:

http://www.pathologyinformatics.org/Resources/neocl.txt

U.S. CENSUS

The U.S. Census can be accessed at: *http://www.census.gov/*

Public datasets can be downloaded from:

http://www.census.gov/popest/datasets.html

Most of the datasets are in simple comma-delimited ASCII format, with a key provided that lists the order and named data elements in each row (record) of the file.

Glossary

$&—Same as "$MATCH". $& represents the part of a string that successfully matches the entire Regex expression:

```
#!/usr/local/bin/perl
$string = "I will arrive in 6 days";
$string =~ /l{2} ?ar[ \w]+[0-9][ \w]+d/;
print $&;
exit;
returns "ll arrive in 6 d"
```

$`—Same as "$PREMATCH". $` represents the part of a string that precedes the string fragment that matches the entire Regex expression:

```
#!/usr/local/bin/perl
$string = "I will arrive in 6 days";
$string =~ /l{2} ?ar[ \w]+[0-9][ \w]+d/;
print $`;
exit;
returns "I wi"
```

$'—Same as "$POSTMATCH". $' represents the part of a string that follows the string fragment that matches the entire Regex expression:

```
#!/usr/local/bin/perl
$string = "I will arrive in 6 days";
$string =~ /l{2} ?ar[ \w]+[0-9][ \w]+d/;
print $';
exit;
returns "ays"
```

`$.`—Same as "`$INPUT_LINE_NUMBER`". As Perl reads sequential lines through a file, it keeps count and stores the current line number in the variable *`$.`*:

```
#!/usr/local/bin/perl
open(TEXT, "myfile.txt");
foreach (1...100)
  {
  $line = <TEXT>;
  }
print $.;
exit;
returns 100
```

`$/`—Same as "`$INPUT_RECORD_SEPARATOR`". In Perl, the end of a line-read operation is determined by the value of the built-in variable *`$/`*. The default value of `$/` is the newline character, `\n`. However, Perl lets you change the value of `$/` to anything you like. This is particularly useful when records in a file are demarcated by a particular string. You can slurp an entire text file into a single variable (assuming your computer has sufficient memory for the task) by simply undefining `$/`:

```
undef($/);
$string = <TEXT>;
```

Perl will read the entire contents of the text file, specified with the file handle `TEXT`, into the variable *`$string`*.

`$1`—When a string matches a Regex expression, it is sometimes useful to know the component of the matching string that matched a particular part of the regular expression. Perl counts the parenthesized parts of a pattern and assigns the matching string component to variables labeled *$1*, *$2*, *$3*, etc., starting from the first parenthesized expression:

```
#!/usr/local/bin/perl
$string = "I will arrive in 6 days";
$string =~ /(l{2} ?a)r[ \w]+([0-9])[ \w]+(d)/;
print "$1\, $2\, $3";
exit;
returns "ll a, 6, d"
```

`$INPUT_LINE_NUMBER`—Same as "`$.`".

`$INPUT_RECORD_SEPARATOR`—Same as "`$/`".

`$MATCH`—Same as "`$&`".

`$POSTMATCH`—Same as "`$'`".

`$PREMATCH`—Same as "`` $` ``".

@ARGV—The list of command line arguments passed to the script.

@_—The list of arguments passed to a subroutine.

abandonware—Software that was once shown to serve some useful purpose but which is no longer used. Almost all of the useful software ever written is now abandonware. In many cases, software loses its value if it is not continually debugged, enhanced, and aggressively marketed. The term abandonware is often applied to software created by graduate students as part of a funded research effort. Once the funding period ends and the students have moved to new pursuits, there is no support for marketing and distributing the software. Even when interest remains in the original software, it can be very difficult for a new programmer to understand the source code left by the original programmer. In addition, funding is typically rewarded for innovative ideas, and it is sometimes impossible to attract new funding to maintain a previously funded project.

acronym—An acronym is a specialized form of abbreviation in which selected letters from a word phrase (usually the first character of each word in the phrase) are sequentially fused to form a new word. See backronym.

ANSI (American National Standards Institute)—ANSI is a standards activities organization. As such, ANSI does not itself develop standards. ANSI accredits standards-developing organizations to create American National Standards (ANS). When something is described as an ANSI standard, it means that ANSI-accredited standards-development organizations followed ANSI procedures and received accreditation from ANSI that all the procedures were followed. ANSI works with over 270 ANSI-accredited standards developers. Groups of ANSI-accredited standards-development organizations work together to develop voluntary national consensus standards and American National Standards (ANS). ANSI coordinates the effort to gain international standards certification from the ISO or IEC.

archived tissue blocks—All tissues removed from patients (for example, by surgeons during operations, by dermatologists who sample small skin lesions, by phlebotomists who draw blood, by patients themselves when they collect urine specimens) go to the pathology department where they are examined. In many cases, samples of tissues are fixed in formalin and then processed to produce a paraffin-infiltrated tissue encased in a block of paraffin. These blocks, sometimes referred to as cassettes (because plastic cassettes hold the paraffin block), are used as the source of thin tissue sections that can be mounted and stained on glass slides. Pathologists look at these glass slides under the microscope and reach a diagnosis based on a correlation of clinical, gross, and microscopic features of the lesions. Unlike radiologists, who look at a visual representations of physical lesions, pathologists look at the actual cells taken from the patient. In most cases, after some of the block has been used to produce tissue sections mounted on glass slides, the majority of the embedded tissue remains in the block and serves as a permanent sample of the biopsied tissue. Some laboratories have paraffin-embedded blocks that have been saved for over a century. Provided that the tissues were properly fixed after they were removed from the patient, these ancient paraffin-embedded tissues are perfectly suitable material for modern research studies. About 25 million surgical pathology specimens are collected each year in the United States. Over the past century, a distributed archive of human tissue blocks has accumulated that has enormous value for biomedical researchers.

array (Perl)—An ordered list of items. In Perl, when a list of items is assigned an array data structure, it gains access to a variety of useful commands that operate on arrays (including `shift`, `unshift`, `push`, `pull`, `join`, `sort`, etc.). The items in a Perl array can be other Perl data structures. It is not uncommon in Perl to have arrays of strings, arrays of arrays, or arrays of hashes.

artificial intelligence (AI)—It is impossible to discuss futuristic concepts noted in this book without some mention of artificial intelligence and so-called "thinking machines." In the opinion of some, formal knowledge specifications such as RDF, DAML, and OWL provide the ingredients for autonomous software agents that perform intelligently and make complex decisions. I am of the opinion that biomedical professionals will spend the next decade collecting, annotating, and merging large heterogeneous datasets. The methods for analyzing such data are still under development. Biomedical ontologies will have simple but important purposes, such as detecting errors in medication, laboratory reports, and physician orders. The field of artificial intelligence in biomedicine still requires a lot of groundwork.

ASCII standard—The American Standard Code for Information Interchange, ISO-14962-1997. The ASCII standard is a way to assign specific 8-bit strings (a string of 0s and 1s of length 8) to the alphanumeric characters and punctuation. Uppercase letters are assigned a string of 0s and 1s that is different from their matching lowercase letters. There are 256 ways of combining 0s and 1s in strings of length 8, so this means there are 256 different ASCII characters. For some uses, the 256 ASCII character limit is too constraining. There are many languages in the world, with their own alphabets or with their own accented versions of the ASCII Romanized alphabet. Consequently, a new character code (UNICODE) has been designed as an expansion of ASCII. To maintain facile software conversion (from ASCII to Unicode) ASCII code is embedded in the Unicode standard.

ASCII editor—See text editor.

assignment (Perl)—In Perl, a variable can be thought of as a container for data. It is convenient to think of assignment as an instruction to put specified data into a named variable:

```
$holder = 5;
```

The value 5 is assigned to the `$holder` variable. The assignment operator is the "=" character, which should not be confused with the "==" operator, which tests the left and right sides of the operator for equality.

associative array (Perl)—See `hash`.

autocoder—A software program capable of parsing large collections of medical records (for example, radiology reports, surgical pathology reports, autopsy reports, admission notes, discharge notes, operating room notes, medical administrative e-mails, memoranda, manuscripts, etc.), capturing the medical concepts contained in the text and assigning them an identifying code from a nomenclature.

backronym—An acronym assigns a shortened form, usually composed of the ordered sequence of the first character of each word in a phrase. A backronym does the opposite by taking the letters in a root word and assigning a sequence of words with each word in the sequence beginning with the ordered letters from the root word. Perl is a word that has backronyms. Various expansions attributed to the letters in Perl were invented after the programming language was named. Practical Extraction and Report Language is a popular backronym for Perl. A less gracious backronym is Pathologically Eclectic Rubbish Lister.

binary data—Technically, all digital information is coded as binary data. Strings of 0s and 1s are the fundamental units of all electronic information. However, when people use the term binary data, they most often are referring to digital information that is not intended to be machine-interpreted as alphanumeric characters (text). Images, sound files, and movie files are almost always binary data files. So-called plain-text files, html files, and XML files (consisting entirely of text characters) are distinguished from binary data files and referred to as plain-text or ASCII files (see ASCII). Confusion arises in files that represent text but which have their own proprietary format for representing text characters and text display instructions. Though these files are intended to display text, they are usually referred to as binary word-processing files.

binmode (Perl)—The binmode command forces files to be read in so-called binary mode, in which variant forms of line delimiters are simply ignored. This command is used on non-UNIX systems and is sometimes necessary to ensure cross-platform compatibility in instances when files are opened and closed by a Perl application:

```
open (TEST, "new.jpg");
binmode TEST;
```

biocurator—A person who collects and organizes biological information relevant to a particular biological field so that it can be effectively understood and used by experimental scientists who work in that field.

biomedical informatics—Uses the data produced by research laboratories (sometimes called discovery data) and the data obtained from clinical repositories to obtain clinically useful results (for example, new discoveries, tests, therapies, services, or procedures). Because biomedical informatics translates basic science into clinical reality, it is typically regarded as a translational or applied science.

Bioperl—In 2000, a group of Perl programmers and bioinformaticians released the first version of the open source BioPerl project. BioPerl provides an interface to public genomic datafaces and implements a variety of algorithms relevant to sequence analysis. The Bioperl Web site is *www.Bioperl.org*

CGI (Common Gateway Interface), Perl—CGI scripts are Perl scripts that operate on an Internet server and carry the suffix "`.cgi`" rather than "`.pl`". CGI scripts reside on Internet servers in the cgi-bin subdirectory and accept input from Web clients through the http protocol and return Web pages to the client. A short example script follows:

```
#!/usr/local/bin/perl
print "Content-type: text/html\n\n";
open (TEXT,"top.txt");
$line = " ";
while ($line ne "")
    {
    $line = <TEXT>;
    print $line;
    }
$myquery = $ENV{'QUERY_STRING'};
if ($myquery =~ /12345/)
    {
   print "<br>You have entered the correct password\n";
    }
else
    {
    print "<br>Boo.....Bad Password\n";
```

```
        print "<br><br><br></body></html> \n\n\n\n";
        exit;
        }
    print "<br><br><br></body></html> \n\n\n\n";
    exit;
```

The Web data from the client is transmitted inside the `$ENV{'QUERY_STRING'}` hash element. This script checks if the data passed from the Web is the password, "12345". If so, it returns a Web page with content and format determined in the `.cgi` script.

`chomp` (`Perl`)—Chomp is the safe alternative to chop. Chomp will remove the last character from a string only if it is the newline character. See `chop`.

```
chomp($filename);
```

`chop` (`Perl`)—The chop command removes the last character from a string. It is often used to chop the newline character from a screen input. See `chomp`.

```
chop($filename);
```

`chr` (`Perl`)—The `chr()` function returns the character represented by the ASCII number included in the expression:

```
chr(65); returns A. see ord().
```

`class` (`Perl`)—In Perl, a class is a package that can create objects (instances of the class with specified attributes) and which contains class methods that are available to the objects.

classification—An organization of everything in a domain by hierarchical groups, according to features generalizable to the group members. Four terms with distinctly different meanings have been used interchangeably with classification, leading to considerable confusion among biomedical informaticians. These terms are: identification, discrimination, taxonomy, and ontology.

clinical trial—An experiment using humans to validate the utility of a treatment or a diagnostic test.

clinical validation—See validation.

`close` (`Perl`)—Closes the file associated with the file handle:

```
open(FILE, "taxo.txt"); #opens file "taxo.txt" for reading
close(FILE); #closes the file handle associated with
"taxo.txt"
```

`closedir (Perl)`—Closes a directory:

```
#!/usr/local/bin/perl
opendir (MYDIR, ".") || die ("Can't open directory");
while ($file = readdir (MYDIR))
  {
  print "$file\n" if (-f $file && -T $file);
  }
closedir (MYFILE);
exit;
```

cluster analysis—A mathematical method of grouping members of a population into clusters based on their similarity.

Common Rule—In the United States, the Common Rule (Title 45 Code of Federal Regulations, Part 46, Protection of Human Subjects) specifies a range of protections for human subjects involved in medical research. The Common Rule specifies that research using medical records is a type of human subject research and falls under Common Rule protection. Under the Common Rule, medical records can be used freely for research purposes if records are made anonymous by removing any identifying links between the record and the patient.

concordance—A complete listing of all the words in a text, along with the locations in the text they occur.

`cos (Perl)`—Perl's built-in cosine function. Perl's cosine function requires radians (not degrees). For example:

```
$rads = (atan2(1,1) * $degrees) / 45;
$cosine_value = cos($rads);
```

The first line converts degrees to radians. Remember that `atan2(1,1)` is pi divided by 4. The second line implements the `cos()` function.

CPAN (Comprehensive Perl Archive Network—A resource for thousands of free Perl modules:

http://www.cpan.org/ports/

curator—Curator derives from the Latin *curatus*, the same root for curative, and conveys that curators take care of things. The curator must ensure that nomenclatures are comprehensive for their knowledge domain, and this can be quite difficult when the knowledge domain is growing rapidly. The modern curator must also ensure that the nomenclature is formatted and annotated in a manner that permits interoperability with other databases. See biocurator.

`curly brackets (Perl)`—In Perl, curly brackets, "{}", are used to begin and end a block of code. Also called curly braces.

DAML (DARPA Agent Markup Language)—Shortly after RDF was invented, it was obvious that it contained no datatype descriptors (that is, formal restrictions on the type of data that can be contained in a data object). RDF also lacked methods for describing the relationships between RDF domains. DAML is an RDF extension written in RDF. DAML is easy to understand and implement within RDF documents:

http://www.daml.org/about.html

DARPA (Defense Advanced Research Projects Agency)—DARPA is run by the U.S. Department of Defense and its intent is to develop technologies of military value. It is surprising that many DARPA efforts have yielded enormous benefit to society. In particular, DARPA has contributed to the design and implementation of the Internet. DARPA's DAML project adds enormous utility to RDF, the key ingredient of the so-called semantic Web.

data annotation—The act of supplementing individual data elements' descriptive data (metadata) and related data from external information sources (for example, clinical or pathological details) for the purpose of enhancing the utility of the data element. Data annotation is usually accomplished with the help of XML (eXtensible Markup Language, see entry in Glossary).

data de-identification—Large medical centers collect terabytes of clinical data every week, but none of this data is directly accessible for research purposes. To obtain aggregated records for the purpose of conducting medical research, a researcher must submit research plans to an Institutional Review Board and/or Privacy Board. Institutional Review Boards are designed to protect patients from harms that might be associated with medical research. In the case of research that uses pre-existing patient records, the risks to the patient are confined to issues of confidentiality or privacy. The risks to the institution are violations of federal or state regulations that restrict the uses of patient records for research. The two federal laws that apply to this situation are the so-called Common Rule and the HIPAA Privacy Regulations. Both of these rules permit the unrestricted use of patient records when the records are de-identified (that is, when the data in the record are disengaged from any links to the patient). There is an intense and urgent interest in developing technical solutions (algorithms, software applications, standard protocols) that permit researchers to create and use large numbers of de-identified medical records.

data integration—Occurs when information is gathered from multiple datasets, relating diverse types of data. Data integration is particularly important to biomedical researchers because data obtained from experiments on human tissue specimens has little applied value unless it can be combined with medical data (that is, pathological and clinical information). In the past, research data was correlated with medical data by manually retrieving, reading, assembling, and abstracting patient charts, pathology reports, radiology reports, and the results of special tests and procedures. Manual annotation of research data is not feasible when experiments involve hundreds or thousands of tissue specimens, so the development of automatic methods for data annotation is an important area of research.

data-intensive biomedicine—Modern biomedical science is data-intensive. Gene, protein, and tissue microarrays allow us to observe changes in thousands of variables, all at once. Vast amounts of data are generated from single experiments. Likewise, large hospital information systems routinely collect and store terabytes of patient-related data. A human biopsy sample used in a high-throughput array experiment is likely to have a wealth of clinical information stored in one or more hospital databases. Each biopsy has a surgical pathology report describing the specimen and listing pathologic findings. The surgical pathology report might contain an archived image of the lesion and a variety of special ancillary tests, including immunohistochemistry and cytogenetics findings. The biopsy used in the array experiment might be one of many different biopsies excised from the same patient, and these non-sampled biopsies might contain information pertinent to the experimental study. The patient's entire medical record, with demographic information (age, ethnicity, gender), history, physical examination, treatment, and outcome might reside in one or more hospital databases. The connected informational resources for an experiment can easily involve terabytes of data.

data mining—A special kind of data searching in which a valued class of data is sought from within one or more data sets. The role of the biomedical data miner is to confer sense upon an otherwise inchoate data collection.

data object—In the simpler past, it was easy to think of data as a number or as a collection of numbers because databases imposed constraints on records and elements. Each record had a specific set of elements, and each element had a constrained datatype. Almost all information can now be represented digitally, and any type of information can be annotated with other information. For instance, an EKG trace can be represented by a binary image, and the binary image can be annotated with information related to the format of the image, the EKG pattern of the image, the diagnosis rendered, a patient identifier, an EKG identifier, etc. The data related to the EKG can be attached to the patient's EMR (electronic medical record) in whole or in part, and can also be attached to other reports generated in the hospital (for example, billing records). Rather than thinking of the EKG as a record, it might be advantageous to think of the EKG as a data object, with a set of possible data elements belonging to the object, as well as a set of methods (software routines) that can be called upon to port sets of these data elements to other objects, and a set of general properties (list of methods) that are shared with other designated data objects. See also object-oriented programming.

data sharing—Data sharing involves one entity allowing data (whole datasets, specific records, or specific items from one or more records) to be accessed by another entity. This process may permit no-charge open access to the public; it can be performed on a fee basis, through contracts and other business arrangements; it might be performed to comply with administrative or regulatory requirements; or it might be performed under the duress of a subpoena. In all cases, medical data-sharing transactions must protect patients from harm (that is, in particular, loss of privacy or breach of confidentiality) and institutions from liability.

De-identification (versus anonymization)—Removal of personal identifiers from text. Closely related to anonymization, but differs in one aspect. De-identification can be reversible. A de-identified document may be linked to another document that permits some of the identifiers to be restored, under carefully specified conditions. An anonymized document must be prepared so that information can never be linked back to individuals.

Developmental Lineage Classification and Taxonomy of Neoplasms—The shortened name is the Neoplasm Classification. This is a free nomenclature that consists exclusively of over 145,000 names of neoplasms. The neoplasms are classified by developmental lineage. Several versions of the classification files are available (XML version, simple ASCII list of neoplasms, and flat-file list of lineage ancestries for each term). See Appendix.

dictionary—A vocabulary or a nomenclature with definitions for each entry.

dictionary (Perl)—A synonym for associative array or hash.

`die (Perl)`—`die` is like `exit`, only with a flourish. The `die` statement in a script ends the program and sends a message to the monitor:

```
(1 == 2) or die "Curse you, Bertrand Russell\!";
```

If 1 does not equal 2 (and it usually does not), Perl ends the script and sends a personal curse to the screen.

differentiated software—Software developers sometimes speak of undifferentiated software and of differentiated software. Differentiated software is developed for a specific use and is often commissioned by a specific user. An example would be a hospital information system designed to meet the needs of a specific facility and to interface with the equipment held at the facility. The same vendor's hospital information system might be sold to many other medical centers, but each local implementation of the software would require additional work and re-engineering of the basic software product.

discrimination—Finding features that separate members of a group according to expected variations in group behavior. Examples of discrimination in medicine are "grading and staging." Grading and staging involve reporting additional morphologic features (grading) or clinical behavior (staging) that place a tumor into a group with a predictable clinical course or response to therapy.

Dublin Core Metadata—There is a fundamental difference between creating a document for oneself and creating a document for others. If you have created a document for yourself, you do not need to include, for example, the name of the person who made the document, the date the document was created, the purpose of the document, or restrictions of the use of the document. But if you have made a document that can be obtained and used by anyone, you must include all this information and more. Librarians understand the importance of having a set of information attached to every document that can be used to index documents for retrieval and to establish the terms under which the document can be distributed and used. The Dublin Core is a set of metadata elements (XML tags) developed by a group of librarians who met in Dublin, Ohio. Every XML document should contain the standard set of annotations for the Dublin Core Metadata Element set (74), (75).

`each (Perl)`—The `each()` function extracts all key/value pairs from a Perl hash. The syntax is demonstrated:

```
while (($key, $value) = each(%example_hash))
   {
   print "$key => $value\n";
}
```

The `each()` function is related to the `keys()` and `values()` functions, which extract all of the keys or values, respectively, and place them into an array. The `each()` function is safe to use in scripts because it can iterate through a hash of any size without creating a data structure to hold the aggregate product. The `keys()` and `values()` functions require sufficient memory to put all of the keys or values into an array.

EHR (Electronic Health Record)—Used interchangeably with EMR.

Electronic Laboratory Notebook (ELN)—ELNs are software applications that allow you to collect data in an organized manner that documents what you have done and when you did it (the kind of information that any experimentalist must produce on demand). ELNs can be used to standardize protocols, manage data quality, establish a convincing data trail, and facilitate data sharing between users. Most ELNs are proprietary software products. Despite claims to the contrary, ELNs are not particularly easy to master. Much of the functionality found in ELNs can be achieved by using common data specifications and an open source programming language to implement data capture and analysis (for example, Perl).

EMR (Electronic Medical Record)—A computerized version of the venerable patient chart. EMR hospital record systems attach all records and data for a patient to his or her EMR data object. Non-EMR hospital information systems tended to allow patient records to reside unconnected in different software modules and applications throughout the medical system. For instance, a patient's dermatology records might have been saved in the dermatology clinic's software package, and this package might have had no connection to the radiologic records kept in the radiology department's PACS system. In theory, an EMR-based system allows a complete set of medical records for a patient to be exchanged between medical centers.

Entrez—The National Library of Medicine's text-based search and retrieval service that includes PubMed, Nucleotide and Protein Sequences, Protein Structures, Complete Genomes, Taxonomy, and other NCBI databases. An Entrez tutorial is available at:

http://www.ncbi.nlm.nih.gov/entrez/query/static/help/entrez_tutorial_BIB.pdf

`environment variables (Perl)`—Data pertaining to your operating system, computer architecture, and CPU that the Perl interpreter stores in a special hash variable called *%ENV*. You can access any or all of the environment variables from a Perl script:

```
while (($key,$value) = each %ENV)
  {
  print "$key=$value\n";
  }
```

Some of the available environment variables are: *homedrive*, *temp*, *systemdrive*, *windir*, *os*, *username*, *path*, and *processor_architecture*. Knowledge of the environment variables can be used in a program to customize functionality for the specific computer on which the application is installed. Unfortunately, the same knowledge can be used to create Perl applications that will not function on certain types of computers.

`eof (Perl)`—Tests for the end of a file. Returns true if the file is at the end:

```
open (TEST, "newfile.txt"); #creates a new and empty file
print eof(TEXT);
returns 1
```

error trap—Perl scripts might fail for reasons unrelated to buggy code. A script might require an external file that is not provided or that cannot be opened because it is in use by another application. A program might require a condition that might not apply (for example, a variable whose content does not exceed a certain maximum number). In these cases, the programmer might prefer the script to exit gracefully or otherwise change its behavior, rather than thrash onward in a doomed attempt to produce a useful output with inadequate input. Perl contains several clever error-trapping constructs, including `carp()` and `confess()`. These devices are analogous to try-catch statements in Java. Well-written Perl scripts contain numerous error traps, but the short scripts in this book lack error traps. One exception is the "die" command used in conjunction with the `open()` function:

```
open (TEXT, $filename)||die"Cannot open file";
```

This command tells Perl to exit the script if the file cannot be opened, and to send an informational message, "Cannot open file," to the monitor.

`escape (Perl)`—In Perl, some characters serve as operators. For instance, the dot means "any character other than a newline" when it appears inside a pattern and it serves as a concatenation operator when it appears outside a pattern. Escaping in Perl is a syntactic method that allows a character to "escape" from its Perlish function so that it can serve as a simple ASCII character. Escaping is accomplished by preceding the character with a slash (for example, "\."). In the following code, the ".", "-", ":", and "_" are escaped:

```
if ($value =~ /^[a-z\_][a-z0-9\-\.\_]*[\:]?[a-z0-9\-\.\_]*$/i)
```

`exit (Perl)`—Ends the script without producing any message, for example, `exit;`.

`exp (Perl)`—Determines *e* to the power of the number provided:

```
print exp 0.99999;
returns 2.71825464577667, just a smidgen less than the
   value of e.
```

`Fcntl (Perl)`—Sometimes Perl needs to extend itself to properly interface with operating systems. The `Fcntl` module provides file controls not native to Perl and is needed in certain types of scripts (for example, SDBM operations). This snippet was extracted from `persist.pl` (List 1.23.2):

```
use Fcntl;
use SDBM_File;
tie%item, "SDBM_File", 'mesh', O_RDWR|O_CREAT|O_EXCL, 0644;
```

`O_RDWR`, `O_CREAT`, and `O_EXCL` are constants provided by the `Fcntl` module that will be understood by the host system.

`for (Perl)`—A `for` loop repeats a block of code while a changing condition evaluates to True. For example:

```
for ($count = 1; $count <= 20; $count++)
```

This sets the index to one, and increments the value of the index by 1 with each iteration of the block, stopping when the index exceeds 20.

`foreach (Perl)`—The `foreach` operator loops through a block of code, assigning to each loop block the next item from an array:

```
foreach $thing (@wordarray)
  {
  print "$thing\n";
  }
```

This snippet prints the items from the `@wordarray` array.

free software—The concept of free software, as popularized by the Free Software Foundation, refers to software that can be used freely, without restriction, and does not necessarily relate to the actual cost of the software. The generally acknowledged father of the free software movement is Richard Stallman, an MIT visionary who led an energetic and unwavering campaign to create and freely distribute some of the most valued software applications in use. The free software movement is similar to the open source software movement, but some of the features of free software (ability to modify and redistribute software in a prescribed manner as discussed in the software license) are not always guaranteed in open source software.

free software license—Virtually all free software is distributed under a license that assigns copyright to the software creator and protects the creator from damages that might result from using the software. Software sponsored by the Free Software Foundation, and much of the software described as either free software or open source software, is distributed under one of the GNU software licenses.

Free Software Movement v Open Source Initiative—Most informaticians use the two monikers interchangeably and do not seem to suffer for the oversight. In practice, there is very little difference between free software and open source software. The most important feature common to free and open source software is that they are non-proprietary.

Richard Stallman has written an essay that summarizes the two different approaches to creating free software and open source software (62).

FOSS (Free and open source software)—This term encompasses both free and open source software and can be used to placate sticklers who distinguish between the two related movements. Another term sometimes encountered is FLOSS (Free Libre Open Source Software), which has international cachet.

FTP (File Transfer Protocol)—An Internet protocol for transferring files between different networked computers. Files can be uploaded or downloaded.

gene expression array—Also known as gene chips, DNA microarrays, or DNA chips. These consist of thousands of small samples of DNA arrayed onto a support material (usually a glass slide). Each sample of DNA is prepared by copying molecules of RNA of known sequence into fluor-tagged sequences of complementary DNA, which are carefully placed on the array. When the array is incubated with cell samples, hybridization will occur between molecules on the array and single stranded complementary (that is, identically sequenced) molecules present in the cell sample. The greater the concentration of complementary molecules in the cell sample, the greater the number of fluor-tagged hybridized molecules in the array. A specialized instrument prepares an image of the array, and quantifies the fluorescence in each array spot. The greater the amount of fluorescence for each array spot, the greater the concentration of the molecule in the cell sample. The dataset, consisting of the fluorescent intensity of each post-hybridization array spot, produces a gene expression profile characteristic of the cell sample. By comparing individual gene expression levels in normal cells and diseased cells, researchers can discover candidate genes that might play a causal role in disease processes. By comparing whole profiles of different tissue samples, researchers might identify subsets of genes representing cellular pathways that distinguish one cell condition from another.

`glob (Perl)`—A non-Perl system command that can be called from Perl. Returns the filenames that match the parenthesized expression, where "*" is a wild-card character:

```
@filelist =  glob('*.xml');
```

This command creates an array of all the filenames in the current directory that end with the filename suffix "`xml`". The same functionality can be achieved using the `readdir` Perl operator.

`gmtime (Perl)`—`Gmtime` converts the value returned by Perl's time function (the number of elapsed seconds since January 1, 1970) into an array of items that can be collected and put into a preferred format that specifies Greenwich Mean Time. The following script converts the items returned by gmtime into the Greenwich Mean Time expressed as the ISO8601 standard time format:

```
#!/usr/local/bin/perl
#ISO8601 time is expressed as YYYY-MM-DD hh:mm:ss, for
#Greenwich Mean Time
@components = gmtime(time());
$year = $components[5] + 1900;
$month = substr(("000" . $components[4]), -2, 2) +1;
$day = substr(("000" . $components[3]), -2, 2);
$hour = substr(("000" . $components[2]), -2, 2);
$minute = substr(("000" . $components[1]), -2, 2);
$second = substr(("000" . $components[0]), -2, 2);
print "The Time is $year\-$month\-$day $hour\:$minute\
   :$second\n";
exit;
returns the time in the form: The Time is 2006-6-1 15:02:03
```

GNU software licenses—The GNU organization publishes two licenses used for software produced by GNU and by anyone who would like to distribute their software under the terms of the GNU license. They are referred to as copyleft licenses, because they primarily serve the software users, rather than the software creators. One GNU license, the General Public License, covers most software applications. The GNU Lesser General Public License, formerly known as the GNU Library General Public License, is intended for use with software libraries or unified collections of files comprising a complex application, language, or other body of work (76).

grant—A grant is an award of money to a researcher in the hope that he or she will work productively with the funds. Most grants are awarded after reviewing a grant application, and in this case, the funders expect that the researcher will work on the project described in the application. The grantee's only real responsibility is to use the grant funds to support their work. The most important distinction between a research grant and a research contract is that the contract specifies a deliverable product, while the grant does not.

`grep (Perl)`—`Grep` is a remarkable tool for packing many Perl operations into a single command line. `Grep` evaluates a Perl expression for each element of an array (locally setting `$_` to each element). If the expression evaluates to true, the result of the expressed operation is added to an output array:

```
@newarray = grep((exists $wordhash{$_}),@lastarray);
```

The preceding code determines whether each of the array items in @lastarray exists as a hash key of %wordhash. If so, the element is added to @newarray:

```
#!/usr/local/bin/perl
opendir(THISDIR, ".");
@filelist = grep (/\.pl$/, grep(-f && -T, readdir(THISDIR)));
print join("\n", @filelist);
end;
```

The preceding code opens the current directory (which operating systems recognize as "."), and reads the directory into an implied array. Grep evaluates the implied array of directory items and determines whether each item is both a file and is composed of text. Items that pass the test are put into another implied array, which is passed to yet another grep operator, which tests whether each of the text files in the implied array has a .pl suffix. Those files with a .pl suffix are put into the @filelist array and printed to the monitor screen using the join() operator to display each array item separated by a newline character.

GRID computing—A method for providing Web services by brokering client requests through a network of participating servers. In the most advanced grid computing architecture, requests can be broken into computational tasks that are processed in parallel on multiple computers and transparently (from the client's perspective) assembled and returned.

halting a Perl script—The most common problem encountered by programmers is non-execution (your script won't run). But another problem, which can be much worse, occurs when your program never wants to stop. This can occur when a block or loop has no exit condition. If you notice that a script seems to run too long, and you want to exit, just press the Ctrl-Break keys (simultaneously). The script should stop and give you back your system prompt. This will not harm your script or your operating system. Sometimes the script is too busy to stop immediately. Just keep pressing the Ctrl and Break keys (simultaneously) a few more times. Eventually, Perl will notice that you want to stop execution.

hash (Perl)—A data structure composed of key/value pairs. Synonyms are Associative Array and Dictionary. Perl hashes are prefixed with the "%" sign. An example is:

```
%dictionary = ("body" => "document","h1" => "title",
   "ul" => "list","ol" => "list");
```

In this example, body and document is a key value pair. h1 and title is another key/value pair. A value can be extracted by invoking its key. For example, the value of *$dictionary{"body"}* is "document." Hashes are probably the most versatile data structures in Perl. They are used in almost every Perl script to store and retrieve data. A common newbie mistake is forgetting that one value can be assigned to many keys, but that each key can have only one value.

heterogeneous data—A pathology report and a histopathology image are very different types of data. One is a narrative text containing diagnostic terms along with clinical, demographic, and administrative information. The other is a binary file relating color and position to pixels. These examples of heterogeneous data are tied by a medical relationship. A digital representation of a bone x-ray might have a relationship to an operative note, a bill for hospital services, or to an accident report. In these instances, the data are likely to have very different types of representations and formats and might have relationships that were not anticipated by the persons who created the data formats for each type of collected information.

HIPAA Standards for Privacy of Individually Identifiable Health Information, Final Rule—Usually referred to under the broader act, the Health Insurance Portability and Accountability Act. Researchers who want to use confidential medical records must fully comply with HIPAA and the Common Rule (see Glossary entry).

HIS (Hospital Information System)—The data system used to collect, store, query and distribute hospital data.

HTTP (Hyper Text Transfer Protocol)—The familiar World Wide Web protocol that permits Web pages to be exchanged over the Internet. Enhancements of server and client functionality (through CGI scripts on the server side and HTML-embedded programs on the client side) empower users to distribute computational tasks using the HTTP protocol.

HL7 (Health Level 7)—HL7 is an international data-exchange specification for medical record data. The term "HL7" applies both to the data exchange specification and to the international group of health standards experts who work on the specification. The HL7 specification provides interoperability between the different devices and record types within a hospital and between different hospitals. For example, it may be used to share records between a laboratory information system (with blood test data) and the hospital's billing system. The latest version of HL7, version 3, permits data to be exchanged through XML, using an HL7 metadata vocabulary.

HMAC (Hashed Message Authentication Codes)—See one-way hash.

information technology (IT)—All of the software and hardware and intellectual property involved in handling electronic data.

`if (Perl)`—`if` executes a block of code when an expressed logical condition evaluates to true. For example:

```
if($my_body eq "dead")
  {
  &proceed_over;
  }
```

`input record separator (Perl)`—"`$/`". The input record separator, newline by default.

`int (Perl)`—The `int` command returns the part of a number that precedes the decimal point:

```
print int(34.7765899);
returns 34.
```

IRB (Institutional Review Board)—IRBs are committees created under the Common Rule by hospitals and research organizations to ensure that human subject research is performed in a manner that protects human subjects. The most important activities of these committees in the realm of data sharing involve ensuring the confidentiality and privacy of patient data.

ISO/IEC 11179 (ISO11179)—XML data is flanked by metadata tags that describe the data. For instance,

<date>October 1, 2005</date>. The metadata in this example is "date." But how do we know what the metadata "date" actually means? How do we know that it refers to a day of the year and not a type of dried fruit? Even metadata needs to be defined, and the ISO/IEC 11179 sets a standard for describing metadata. Well-defined metadata should be defined in an accessible document that includes the following information for the metadata tag:

- name—the label assigned to the tag
- identifier—the unique identifier assigned to the tag
- version—the version of the tag
- registration authority—the entity authorized to register the tag
- language—the language in which the tag is specified
- definition—a statement that clearly represents the concept and essential nature of the tag
- obligation—indicates if the tag is required to always or sometimes be present (contain a value)
- datatype—indicates the type of data that can be represented in the value of the tag
- maximum occurrence—indicates any limit to the repeatability of the tag
- comment—a remark concerning the application of the tag

ISO8601—An international standard for expressing date and time. See `gmtime`.

`join (Perl)`—The opposite of split, `join` takes all the members of an array and concatenates them into one scalar, with items separated by a provided character or string:

```
print join("\n",@finalarray);
```

justification—In printing, justification is the process whereby letter spacing and word spacing within a line is adjusted so that each line on a page ends evenly, producing a straight right-side margin. Whenever you read an e-mail with lines that extend beyond the monitor screen or lines abruptly chopped in the middle of the page, you are the innocent victim of an unjustified textual assault.

`keys (Perl)`—Returns all the keys (of the key/value pairs) from a Perl hash, as an array. See `values`.

```
@keysarray = keys(%dictionary);
```

`last (Perl)`—The `last` command exits a loop immediately and proceeds to the next line following the loop:

```
#!/usr/local/bin/perl
while (1)
 {
 last;
 }
print "The end";
exit;
```

leaning toothpick syndrome—Extensive use of forward slashes and back slashes in Perl for different operations within the same command line can produce code that resembles jumbled toothpicks. In the following example, slashes are used to demarcate patterns and to escape nonalphanumeric characters:

```
$string =~ s/\<[^\<]+\>//;
```

`length (Perl)`—Returns the number of characters in a scalar variable:

```
if (length($grepline)>73)
```

`lib (Perl)`—When calling external modules, your Perl script needs to know the directory it should search to find external modules. If your module is located in Perl's default directory, it is not necessary to invoke "use lib", but otherwise, include a "use lib" statement prior to invoking the module. For example:

```
use lib "C:/ftp/MIME-Base64-Perl-1.00/lib";
use MIME::Base64::Perl;
```

`log (Perl)`—Returns the natural logarithm of the provided expression:

```
print log (2.7);
returns 0.993251773010283
```

`loop (Perl)`—A `loop` is a block of code that executes repeatedly while the condition specified at the top of the loop holds true. Examples of loop conditions are `while`, `for`, `foreach`, `until`, and `unless`. In Perl, the loop block is enclosed by curly brackets. Loops can nest other loops.

machine translation—Usually refers to programs that transform free text from one language into another. In the realm of biomedical informatics, the term machine translation is sometimes expanded to include autocoding (extracting terms from free-text that match a coded vocabulary, such as UMLS or MESH).

medical vocabulary—See vocabulary.

MedIS (Medical Information Systems)—Includes any software applications employed in delivering health care. Examples are: HISs, LISs, PACS, imaging systems and image analysis software, and radiation therapy systems.

MESH (Medical Subject Headings)—See Resources.

memoization—A method that reduces the performance decline inherent in recursive subroutines. Memoization stores results of calculations from a recursive subroutine in a hash (or some other data structure). When the recursive subroutine is called, the memoization subroutine "looks" in its hash to see if the recursive calculation has been previously stored. If so, the hash value substitutes for the recursive calculation.

middleware—An ill-defined realm of software development curiously described by middleware guru Ken Klingenstein as the "intersection of the stuff that network engineers don't want to do with the stuff that applications developers don't want to do." In more practical terms, middleware is software that operates over a network and supports interoperation between two different software applications. Closely related terms are mediators, wrappers, and agents. All these software tools function to merge, explain, or intelligently process diverse types of data spread through a network.

minimum necessary principle—An understanding that when medical data are shared for the purpose of conducting research, there is an ethical obligation to share only that portion of the patient record that is actually needed to conduct the research.

`module (Perl)`—Special types of Perl packages. Modules are always stored in a single file, and the module filename is the same name as the package it contains, with the suffix `.pm`. Traditionally, modules begin with an uppercase letter. Modules are used extensively in object-oriented programming, but not all Perl modules support object-oriented programming principles. See:

http://world.std.com/~swmcd/steven/perl/module_anatomy.html

MRCON—Pronounced Mr. Con, MRCON is the file containing every term in the UMLS metathesaurus. Each term is assigned a CUI code (concept unique identifier). The following is an example of the first record in the MRCON file. The online MRCON record is broken here into two lines, to fit the printed page:

C0000005|ENG|X|L0000005|X|S0007492|X|A7755565||M0019694|

D012711|MSH|PEN|D012711|(131)I-Macroaggregated Albumin|0|N||

There are about six million MRCON terms in the 2005 UMLS.

`my (Perl)`—Placed before a variable name, `my` ensures that the variable is local, existing only for the duration of the current subroutine.

natural language processing (NLP)—NLP involves parsing through free text (also called narrative text, written text, prose, and unstructured data) to produce a particular organization, transformation, or annotation that corresponds, in whole or in part, to the intended meaning of the original text. Autocoding can be performed with NLP. The parsed text is tokenized into parts of speech, recombined, and normalized to conform to the typical organization of parts of speech found in a preferred nomenclature, and then it is mapped to the identical or equivalent terms found in the nomenclature. As an editorial comment of no promised merit, the author believes that NLP-based implementations are always slow and inaccurate. NLP processing should be avoided when free text can be computationally analyzed by some other method.

Neoplasm Classification—See Developmental Lineage Classification and Taxonomy of Neoplasms.

newbie—A novice or newcomer. Usually refers to someone who is just beginning to use a programming language or an intricate technology. Newbies should begin their conversations with Perl experts by confessing their newbie status. Experts will often tolerate uninspired questions if they know that they are coming from a newbie.

`next (Perl)`—In a looping block, tells Perl to ignore the remaining block commands and return to the next loop of the block:

```
while ($hold ne "")
  {
  $hold = <TEXT>;
  if ($hold eq "\n")
     {
     print HOLDFILE "\n\n";
     next;
     }
  $hold =~ s/\n/ /g;
  print HOLDFILE $hold;
  }
```

NLP—See natural language processing Glossary entry.

nomenclature—Comprehensive collection of the words contained in a specific knowledge domain. An example of a fine, open access biomedical nomenclature is MESH (Medical Subject Headings):

http://www.nlm.nih.gov/mesh/

notation 3—Also called n3. A shorthand syntax for RDF providing a cleaner, less verbose format. For example, a short script that converts an n3 file into formal RDF syntax:

```
#!/usr/local/bin/perl
open(STDOUT, ">image.rdf");
use RDF::Notation3;
use RDF::Notation3::XML;
$rdf = RDF::Notation3::XML->new();
$rdf->parse_file("image.n3");
$string = $rdf->get_string;
print $string;
exit;
```

object-oriented programming—A special type of programming that combines data structures with methods, creating reusable objects.

off-the-shelf software (OTS software)—A generally available software component, used by a medical device manufacturer for which the device manufacturer cannot claim complete software life cycle control (77). The FDA has written guidelines for the use of OTS software in medical devices:

http://www.fda.gov/cdrh/ode/guidance/585.pdf

one-way hash—An algorithm that transforms a string into another string in such a way that the original string cannot be calculated by operations on the hash value (hence the term "one-way" hash). These popular algorithms are discussed in HIPAA, where they are referred to as HMACs (Hashed Message Authentication Codes). Examples of public domain one-way hash algorithms are MD5 (28) and SHA, the Standard Hash Algorithm (29). These differ from encryption protocols that produce an output that can be decrypted by a second computation on the encrypted string. Synonyms are message digests and secure checksum.

ontology—A rule-based grouping of members of a knowledge domain. Ontologies support queries and logical inferences pertaining to the (ontologic) group members. They can be used to test data for logical consistency and can be designed to discover relationships between different classes of data.

`open (Perl)`—Opens the file associated with the file handle:

```
open(FILE, "taxo.txt");  #opens file "taxo.txt" for reading
close(FILE);
#closes the file handle associated with "taxo.txt"
```

open access—Open access applies to documents in the same manner as open source applies to software. A formal definition of open access is available at:

http://www.biomedcentral.com/openaccess/bethesda/

`opendir (Perl)`—Opens a directory for reading:

```
#!/usr/local/bin/perl
opendir (MYDIR, ".") || die ("Can't open directory");
while ($file = readdir (MYDIR))
  {
  print "$file\n" if (-f $file && -T $file);
  }
closedir (MYFILE);
exit;
```

open source—The open source software movement is an offspring of the free software movement. The open source movement was created, in part, to placate developers who wanted to sell software and felt the term "free" as in "free software movement," would be misconstrued by prospective customers to mean that the developer requires no remuneration. Although a good deal of free software is no-cost software, the intended meaning of the term "free" is that the software can be used without restrictions. The term "open source" obviates the need to draw this distinction. The Open Source Initiative posts an open source definition (78) and a list of approved open source licenses (79).

open standard—How could a standard not be "open"? Sadly, there are no prohibitions against imposing use and distribution restrictions on standards. Many standards are available only as copyrighted documents sold by the organizations that developed the standard. Some standards have restrictions on the distribution of the standard that might extend to all materials annotated with elements of the standard.

`ord (Perl)`—The `ord()` function returns the ASCII value of the first character of an expression:

```
print ord("B");
Prints 66, the ASCII value of "B"
```

P2P (Peer-to-Peer or Peer-2-Peer)—A protocol for distributing messages (queries and responses) through a network of peers. Peers are networked host computers that, for purposes of the P2P protocol, are somewhat autonomous and have properties that are neither client-type nor server-type.

`package (Perl)`—Files or declared portions of files that create namespaces for the variables and subroutines in the package. When the package is declared in the calling script and in the package file, there is privacy for variable and subroutine names.

`package statement (Perl)`—A package will contain a statement that declares the package. For example:

```
Package Biostuff;
```

This statement will typically appear in a file named `Biostuff.pl`. To call the `Biostuff.pl` package from another Perl script, the calling script will typically contain a require statement:

```
require 'Biostuff.pl';
```

`PACS (Picture Archive Communication System)`—Used by radiology departments and some image software applications to store and distribute digital images.

pattern matching (Perl)—See `Regex`.

PDL (Perl Data Language)—PDL is a library of math functions, many of which are devoted to matrix (specialized array) operations. PDL can be used in image processing and in graph plotting. Abundant information on PDL is available from *http://pdl.perl.org*

peer-to-peer —See P2P Glossary entry.

pointers—Pointers in C are equivalent to references in Perl. See references.

polysemy—In polysemy, a single word, character string, or phrase has more than one meaning. Polysemy can be thought of as the opposite of synonymy, wherein different words all mean the same thing. Polysemy is a particularly vexing problem in the realm of medical abbreviations. A single acronym might have literally dozens of possible expansions.

`pop (Perl)`—`pop` shortens the array by one item, and returns the value of the last item to a variable. `pop` is the opposite of `push()`:

```
$oldline = pop(@eighty_line);
```

`print (Perl)`—The `print` command prints the value of a variable to a file, to the screen, or to some other output device. Printing the value of a variable to a file is the same as writing the variable to the file.

proband—Someone with a genetic disease whose disease status has been verified as part of a formal genetic investigation. Also called index case.

public domain—Works in the public domain are not copyrighted and anyone can use, copy, and distribute such works freely. Most data produced by the U.S. government cannot be copyrighted and falls into the public domain. Copyrighted materials that have exceeded the term limits of copyright also fall into the public domain. Some authors simply waive copyright on published materials, sending their works directly to the public domain. Data, documents, and books that are not owned by anyone fall into the public domain. Sometimes people forget to provide proper citation to authors of public domain material. Public domain material must be cited just like any other original information source. To take false credit for another person's effort is plagiarism, even when the material is public domain. Authors who want their works to be freely copied and distributed, but who want to retain some sense of attachment to their works, might consider establishing copyright under the GNU free documentation license. The GNU documentation license is designed for manuscripts and books:

http://www.gnu.org/copyleft/fdl.html

PubMed—PubMed is the National Library of Medicine's public database of medical journal articles. It provides easy search access to millions of journal abstracts:

http://www.pubmed.org

`push (Perl)`—Adds an item to the end of an array. `push` is the opposite of `pop()`:

```
push(@newlinearray,$thing);
```

`qq (Perl)`—The `qq|...|` notation is a Perl cure for the leaning toothpick syndrome. It tells Perl to accept the literal contents of the flanked expression, without needing to backslash special characters:

```
qq|<html><head><title>Something</title>|
```

same as:

```
"\<html\>\<head\>\<title\>Something\<\/title\>"
```

quitting a Perl script—See halting a Perl script.

`qw (Perl)`—The `qw` operator is a handy device that accepts array lists without commas or string quotes: `@elements = qw(gene 4gene gene:ncbi gene-autry ge::ne);`

`rand (Perl)`—Perl's random function returns a nearly random decimal number, between 0 and the value passed to the function. For example:

```
rand(26);
possible output: 12.8563842773438
impossible output: 34
```

The `rand()` function is indispensable for cryptographic work, Monte Carlo simulations, and resampling statistics.

RDF (Resource Description Framework)—A special syntax within XML that constrains content to assertions that consist of a declaration of a specified object followed by a metadata/data pair of information pertaining to the object. These assertion triples (specified object, metadata, data pertaining to the specified object) are all that are necessary to bind data to subject and to create a statement of meaning. These statements of meaning can be aggregated with other statements from the same dataset or from other datasets, so long as they pertain to the same specified object. A simple syntactical ploy, such as RDF, is the foundation for the semantic Web, in which logical inferences can be drawn from meaningful assertions (RDF triples) distributed throughout the Internet.

RDF ontology—Though ontologists might find this definition lacking in rigor, an RDF ontology is simply an RDF schema that consists exclusively of classes (that is, no properties).

RDF schema—A formal dictionary of the classes and properties related to a data object, prepared in a special RDF syntax made for schema elements. RDF documents link to RDF schemas when they create new instances of a class and when they use metadata corresponding to properties described in the RDF schema. An RDF document can refer to zero, one, or many RDF schemas.

`read (Perl)`—`read` is an operation that extracts a specified number of bytes from a file, beginning at a specified file location, and puts the extracted data into a variable. It is almost always used in conjunction with `seek()`. In the following example, the `seek()` command tells Perl to seek the location in the file whose file handle is `TEXT`, starting from the beginning of the file, and stopping at the byte designated by the position $place minus 1000 bytes. The `read()` command tells Perl that starting from the position determined by the `seek()` command, Perl should grab 20000 bytes from the file and put the extracted bytes into the variable, *`$holder`*:

```
seek(TEXT,($place - 10000), 0);
read(TEXT,$holder,20000);
```

`readdir (Perl)`—Reads a directory: Example:

```
#!/usr/local/bin/perl
opendir (MYDIR, ".") || die ("Can't open directory");
while ($file = readdir (MYDIR))
  {
  print "$file\n" if (-f $file && -T $file);
  }
closedir (MYFILE);
exit;
```

`ref (Perl)`—The `ref` function tells you the type of data structure "referred to" by the reference variable. The possible returned values are:

SCALAR

ARRAY

HASH

CODE (if it's a reference to a subroutine)

IO (if it's a reference to a file handle)

GLOB (if it's a type glob)

REGEXP (if it's a reference to a precompiled pattern)

REF (if it's a reference to another reference)

undef (if it's not a reference variable, but just a scalar variable)

name of the Class (if the variable is a class object)

`references (Perl)`—A Perl reference is a variable that points to a memory location in which a Perl data structure is stored. Dereferencing a reference variable yields the data structure. Perl has a specific syntax for creating reference variables and for dereferencing individual items stored in the [referenced] data structures. References are not covered in this book, but readers should know that references in Perl (equivalent to pointers in C) lie at the heart of many Perl operations. In particular, references work behind the scenes to make object-oriented programming possible.

`Regex (Perl)`—`Regex` is a shortened form for regular expression. It is a formal syntax for describing string patterns. The syntax, once mastered, permits programmers to write remarkably clever and useful pattern-matching routines. Many different programming languages and mark-up languages use the same `Regex` syntax. An example of `Regex` is "b[na]+", which represents a string pattern that begins with the letter b and is followed by any one of the letters "n", or "a" with at least one occurrence. So "banana" would match the regular expression, as would "baaaaaa", signifying that the regular expression could be useful to computer programmers, monkeys, or sheep.

`regular expressions (Perl)`—See `Regex`.

`require (Perl)`—The "require" statement loads an external Perl script into another Perl script (the requiring Perl script). It is virtually identical to the `#include` statement in the C language. The syntax for "require" is "require `scriptname.pl`;". The "require" statement is different from Perl's "use" statement in that "use" statements are executed at compile-time, while "require" statements are executed at run-time. In addition, "use" statements accept only module names (that is, Perl scripts named with a `.pm` suffix).

`return (Perl)`—In a subroutine, the "return" statement specifies the variable that will be passed back to the script command line that called the subroutine.

`reverse (Perl)`—Reverses the order of items in an array.

RPC (Remote Procedure Calling)—A protocol that permits a program running on a client computer to invoke a computational method and call another networked computer to execute the method, returning the result as a variable that is used in the calling program. This is all done without interrupting the flow of the calling program. RPCs are the purest form of networked computation.

`s (Perl)`—Incredibly useful `Regex` operator that will find a matching pattern in a string and substitute any string you specify. Has the form s/pattern to match/pattern to replace/. For example:

```
$string =~ s/ [a-z]{1}/uc($&)/ge;
```

The preceding code changes the first letter of lowercase words appearing after a space to uppercase. The `g` modifier tells Perl to repeat for each match occurrence, and the `e` modifier tells Perl to evaluate any code that might be present in the replacement pattern.

scrubbing—The process of removing objectionable works, private information, and personal identifiers from text.

`seek (Perl)`—An operation that takes you to a specified location within a file. It is almost always used in conjunction with `read()`. In the following example, the `seek()` command tells Perl to seek the location in the file whose file handle is `TEXT`, starting from the beginning of the file, and stopping at the byte designated by the position *`$place`* minus 1000 bytes. The `read()` command tells Perl that starting from the position determined by the `seek()` command, Perl should grab 20000 bytes from the file and put the extracted bytes into the variable, *`$holder`*:

```
seek(TEXT,($place - 10000), 0);
read(TEXT,$holder,20000);
```

`seekdir (Perl)`—The `seekdir()` operator acts on directories in the same manner as `seek()` acts on files.

semantics—The study of meaning (Greek root, *semantikos*, significant meaning). Data can be structured and meaningless. Consider the assertion: Sam is tired. This is an adequately structured sentence with a subject, verb, and object. But what is the meaning of the sentence? There are many people named Sam. To which Sam does this sentence refer? What does it mean to say that Sam is tired? Is "tiredness" a constitutive property of Sam, or does it apply only to specific moments? If so, for what moment in time is the assertion "Sam is tired" actually true? To a computer, meaning comes from assertions that have a specific, identified subject associated with some sensible piece of fully described data (metadata coupled with the data it describes). As you might suspect, most data contained in databases is not strictly "meaningful."

`shebang (Perl)`—Perl scripts always start with a shebang, a pound character, "#" (known by the musically inclined as a sharp character), followed by an exclamation sign, "!" (connoting a surprise or a bang). For example:

```
#!/usr/bin/perl
```

`shift (Perl)`—Extracts the first item of an array, shortening the array by one. Opposite of `unshift()`. For example:

```
#!/usr/local/bin/perl
@samplearray = qw(Perl rules the world);
print shift(@samplearray);
exit;
returns Perl
```

`sin (Perl)`—Perl's built-in sine function. Perl's sine function requires radians (not degrees). For example:

```
$rads = (atan2(1,1) * $degrees) / 45;
$sine_value = sin($rads);
```

The first line converts degrees to radians. Remember that `atan2(1,1)` is pi divided by 4. The second line implements the `sin()` function.

SOAP (Simple Object Access Protocol)—An XML protocol for expressing Web service requests and for encapsulating Web service replies, using a syntax that is operating-system and programming-language independent.

`socket (Perl)`—A `socket` is a software device that allows you to make connections through the Internet. The module that implements sockets is IO::Socket.

software agent—A computer program that can operate in a somewhat autonomous fashion: collecting data, making logical inferences, and proceeding based on automated decisions. Though the definition of a software agent varies, most definitions convey the idea that software agents can interact with other software agents. This requires each software agent to contain instructions to describe itself using a standard data format that other agents understand. A special breed of software agent, the autonomous agent, proceeds through multiple interactions without human supervision.

software interoperability—It is unrealistic to expect everyone to use the same operating system, the same programming language, and the same software applications. It is nonetheless often necessary to have software that can operate with other software regardless of differences in operating systems and programming languages. There is a wide variety of methods by which this can be achieved. The topic of software interoperability has become complex, but it remains a fundamental issue in all attempts to share analytic methodologies.

`sort (Perl)`—`sort` is a built-in Perl function that sorts the items of a list in string comparison (that is, ASCII) order. Numbers are treated as strings of ASCII characters and will not sort in numeric order using the unaided sort routine.

specification—A standard way to describe something using well-defined descriptors and well-defined units of measurement and organizing the descriptive data in a manner that can be unambiguously understood.

`split (Perl)`—`split` is one of the most useful Perl functions. `split` chops a scalar into items delimited by a provided string or pattern, and puts the ordered items into an array. `split` is particularly useful for splitting a data record into its component data elements or splitting text into individual words. The following splits text at newline or space characters, effectively putting the sequentially occurring words into an array variable:

```
@wordarray = split(/[\n\s]+/, $all_text);
```

`sqrt (Perl)`—Returns the square root of a number:

```
print sqrt(7);
returns 2.64575131106459
```

standards development organization—There are hundreds, perhaps thousands, of organizations that develop various types of specifications. Standards development organizations might become members of a standards activities organization, such as the American National Standards Institute, so that their specification can become a national or international standard.

standards organizations—There are only a few organizations that certify new standards, and in the field of biomedical informatics, the two most important are ISO (International Organization for Standardization) and IEC (International Electrochemical Commission).

stop words—High-frequency words such as the, and, an, but, and if that tend to delineate phrases or terms in text. Stop words are used in machine translation projects and are also called barrier words.

string—In computer parlance, a string generally refers to a sequence of ASCII characters (that is, a character string). All words are strings. All phrases are strings. All sequences of digits are strings. A variable in Perl typically holds a string.

`sub (Perl)`—In Perl, subroutines start with the word "sub". Perl draws no distinction, as some other languages do, between subroutines and functions:

```
&builder #tells Perl to drop to the builder subroutine;
#returns to this line after subroutine block is completed
sub builder
  {
  #does something here
  }
```

`subroutine (Perl)`—A `subroutine` is a named block of code that receives parameters passed as a list of variables and returns its last evaluated expression to the line that called the subroutine:

```
#!/usr/local/bin/perl
$result = &adder("1","2","3"); #passes three variables to
                                        #subroutine
print $result; #last line of script prints $result
sub adder
  {
foreach $parameter (@_) #The @_ array contains passed
```

```
                                    #variables
        {
        $total = $total + $parameter;
        }
      return $total; #the value of total is returned to $result
      }
    end;
```

`substitution (Perl)`—See `s (Perl)`.

`substr (Perl)`—Extracts a substring of a string (first parameter), beginning from a specified position (second parameter) and extending a specified number of characters (third parameter). If the second parameter is negative, it counts back from the end of the string:

```
print substr("0000000231",-5,5);
returns 00231
```

taxonomy—The list of all the instances in a knowledge domain. Taxonomies are used to fill a classification or an ontology. In the case of cancer, a neoplasms taxonomy would be the complete listing of all the different named tumors.

`tell (Perl)`—`tell()` returns the current byte position in the file:

```
tell(TEXT);  #TEXT is the filehandle;
```

Telnet—A network protocol for establishing command-line sessions on remote host computers.

terminology—See nomenclature.

text editor—A text editor (also called ASCII editor) is a software program designed to display simple unformatted text files. They differ from word-processing software applications that produce files with formatting information embedded within the file. Text editors use the ASCII standard code that converts 8-bit binary sequences into alphanumeric characters. Text editors have certain important qualities that word processors lack. They allow you to inspect all the characters contained in a file. Text editors are much faster than word processors because they simply display the contents of files without having to interpret and execute formatting instructions. They can typically open files of enormous size with ease. Most word-processing files choke on files that are hundreds of megabytes in length. Examples of free and open source text editors are Emacs and vi.

thesaurus—A vocabulary that groups together synonymous terms.

threshold protocol—Cryptographic protocol that divides messages into multiple pieces, with no single piece containing information that can reconstruct the original message.

`tiehash (Perl)`—A hash that is tied to a persistent external database object. Normally, all Perl variables die when the Perl script ends. If a script ties a hash variable to an external database object, the members of the hash can be accessed by any Perl script, without needing to rebuild the original hash variable. Creating the tiehash, *mesh*, for hash variable *`%item`* is done in the following:

```
tie %item, "SDBM_File", 'mesh', O_RDWR|O_CREAT|O_EXCL, 0644;
```

`time (Perl)`—The value returned by Perl's `time()` function is the number of elapsed seconds since January 1, 1970. See `gmtime`.

time-stamp—There are many times when it is important to know when a document was produced. Simply supplying a time and date does not suffice. You could be purposefully wrong or you could provide wrong information inadvertently. Every computer can compute time and date. The file systems of virtually all computer operating systems automatically add time and date information to every created file. Unfortunately, falsely setting the computer's time is a simple matter. Time-stamping sometimes means regular checks of the computer clock against an external standard clock and saving time-stamped transaction data on an external medium that cannot be altered (so that later modifications to transaction times can be detected). In some instances, the use of a time-stamp service may be required.

tissue blocks—See archived tissue blocks.

tissue microarray (TMA)—TMAs, first introduced in 1998, are collections of hundreds of tissue cores arrayed into a single paraffin histology block. Each TMA block can be sectioned and mounted onto glass slides, producing hundreds of nearly identical slides. TMAs permit investigators to use a single slide to conduct controlled studies on large cohorts of tissues, using a small amount of reagent. The source of tissue is only restricted by its availability in paraffin and ranges from cores of embedded cultured cells to tissues from any higher organism. In a typical TMA study, every TMA core is associated with a rich variety of data elements (image, tissue diagnosis, patient demographics, other biomaterial descriptions, and quantified experimental results).

Tissue Microarray Data Exchange Specification—Under ideal circumstances, a single paraffin TMA block can be sectioned into nearly identical glass slides dispensed to many different laboratories. These laboratories might use different experimental protocols. They might capture data using different instruments, different databases, different data architectures, different data elements, and immensely different formats. These laboratories could vastly increase the value of their experimental findings if they could merge their findings with those of the other laboratories that used the same TMA block. A key barrier to this process was the incompatibility between datasets produced in different laboratories. The TMA data exchange specification allows researchers to submit their data to journals and public data repositories and to share or merge data from different laboratories. The TMA data exchange specification is a well-formed XML document with four required sections: 1) header, containing the specification Dublin Core identifiers; 2) block, describing the paraffin-embedded array of tissues, 3) slide, describing the glass slides produced from the block; and 4) core, containing all data related to the individual tissue samples contained in the array. Eighty CDEs, conforming to the ISO-11179 specification for data elements, constitute XML tags used in the TMA data exchange specification. A set of six simple semantic rules describes the complete data exchange specification.

`Tk (Perl)`—A Perl extension that allows users to create GUIs (graphic user interfaces) for their Perl applications. `Tk` is not covered in this book for several reasons: 1) This book is about self-sufficiency, and this means readers are expected to create Perl scripts for their own uses and probably will not choose to include a fancy graphic user interface; and 2) This book is about biomedicine. There's nothing particularly biomedical about GUIs. There are many books available on `Tk`, for instance, *Mastering Perl/Tk* by Stephen O. Lidie and Nancy Walsh (O'Reilly), and *Learning Perl/Tk: Graphical User Interfaces with Perl*, by Nancy Walsh (O'Reilly).

`tr (Perl)`—`tr` is Perl's character translation operator. The operator takes specified characters or ranges of characters and translates them to any other chosen characters:

```
$variable =~ tr/b/c/; #translates every b to a c in $variable
$variable =~ tr/0-9/ /; #translates every digit to a space
$variable =~ tr/A-Z/a-z/; #translates uppercase to lowercase
$variable =~ tr/\:/\./; #replaces colons with periods
```

translational research—Includes all biomedical research that attempts to find new therapies, devices, techniques, or tests that can be used in a clinical setting. Much translational research is aimed at deriving benefit from the many advances in basic biomedical research achieved in the past decade. Data-sharing methodologies would be a type of translational research.

triples—The essence of RDF is that all assertions are composed of three things: a subject, a piece of data (about the subject), and a metadata element (that describes the data element). All RDF documents can be deconstructed as collections of triples. Here is a short script to do just that:

```
#/usr/local/bin/perl
use Data::Dumper;
use RDF::Core::Parser;
my %options = (Assert => \&handleAssert,
```

```
                        BaseURI => ".",
                        BNodePrefix => "rdf");
my $parser = new RDF::Core::Parser(%options);
$parser->parseFile('image.rdf');
sub handleAssert
  {
  print Dumper(@_);
  print "\n\n";
  }
exit;
```

UML (unified modeling language)—UML is a standard methodology, developed by the OMG (object management group) for specifying how a complex software application works. Properly specifying software by using UML notation (or using UML diagramming techniques) requires significant expertise. For simple software scripts (the kind discussed in this book), UML is not necessary.

UMLS (Unified Medical Language System)—See the Appendix.

`undef (Perl)`—`undef` kills variables by undefining them. For example:

```
undef($/);
$variable = <READFILE>;
```

This command undefines the input record separator ("`$/`"). With no defined separator between records, a file line-read will slurp the entire contents of the file (`READFILE`) into an assignment variable (*`$variable`*).

undifferentiated software—Software developers sometimes speak of undifferentiated software and of differentiated software. Undifferentiated software comprises the basic algorithms that everyone uses and re-uses in the process of developing new software applications. All software developers have reasons to keep this basic kind of software free, because nobody knows who developed the algorithms originally and nobody wants to devote their career to prosecuting or defending tenuous legal claims over the ownership of the fundamental building blocks of computer science.

unique identifier—An elusive concept. A unique identifier identifies an object in a manner that distinguishes the object from all other objects. Three conditions should hold: 1) the unique identifier for an object can serve as the name of the object; 2) if two objects have the same unique identifier, they must be equivalent (that is, interchangeable) objects. In most instances (such as identifiers for people), two objects cannot have the same identifier; 3) if an object has more than one unique identifier (for example, an LSID identifier and a DICOM identifier), the object itself must contain, in an accessible form, all the unique identifiers that apply.

The third condition needs some explanation. If an object does not contain all its different unique identifiers, an object can be distinguished from itself. In the case of LSID and DICOM identifiers, the LSID identifier could (erroneously) distinguish the object from itself, identified by the DICOM identifier. There must be a way of knowing that the LSID identifier and the DICOM identifier both represent the same object. The only reliable way to accomplish this is to have both pieces of data encapsulated by the object. Of course, the best approach is to forbid subsequent identifiers once the first identifier has been assigned.

unique object—Objects that have immutable features, which make the object different from all other objects. Every person is a unique object. Every moment in time is a unique object. The original Eiffel Tower in Paris is a unique object. The class of "architectural towers" may not qualify as a unique object, but it can be a fully specified object, and it can be assigned a unique identifier.

`unless (Perl)`—`unless` executes a block of code when an expressed logical condition evaluates to false. It is the opposite of `if()`:

```
unless($my_body ne "dead")
  {
  &proceed_over;
  }
```

Same as:

```
if($my_body eq "dead")
  {
  &proceed_over;
  }
```

`unlink (Perl)`—Deletes a list of files or a single file. For example:

```
unlink "my_file.txt";
```

`unshift (Perl)`—`unshift` takes an item and puts it in front of an ordered array (that is, assigns it the zeroth item, with all other items moved one higher in their array order). For example:

```
unshift(@setlist,$thing);
```

`unshift` is the opposite of `shift`. See `shift`.

`until (Perl)`—Used in loops, setting a condition that the loop continues until a condition is satisfied:

```
until ($line eq "\n")
  {
   $line = <STDIN>;
```

```
        print OUT $line;
        print NEW $line;
        }
    exit;
```

URL (uniform resource locator)—The Internet address of a net resource, such as a Web page or an ftp file.

validation—One of the best uses of biomedical data is to validate assertions using existing data sources. You validate an assertion (which might appear in the form of a hypothesis, a statement about the value of a new laboratory test, or a therapeutic protocol) by showing that you draw the same conclusion repeatedly whenever you analyze relevant datasets. It might be useful to compare reproducibility and validation. Reproducibility is when you get the same result (for example, a diagnostic test result or an observed experimental value) over and over when you perform the test. Validation is when you draw the same conclusion over and over. Biomedical datasets might contain millions of test observations, often annotated with clinical outcome data and epidemiologic data. Many people believe that a wide range of clinically important validation efforts could be performed quickly and inexpensively if the data collected in hospitals were made available to researchers.

`values (Perl)`—Perl hashes are composed of key/value pairs. The values of a Perl hash can be extracted as an array using a `values()` function. The order of array values seems to be random and corresponds to the way Perl chooses to store hash data structures. The `keys()` function would produce the array of hash keys in the same, apparently random, order as the `values()` function. This means that the nth item in the array produced by `keys()` and the nth element in the array produced by `values()` should correspond to a proper key/value pair contained in the original hash. In most instances, you will find that the `each()` operator is the appropriate and preferred function to extract key/value pairs from a hash. See `keys`.

variable—A variable in computer science is somewhat different from a variable in mathematics. You can think of a variable as a named container for a specified piece of data. In Perl, the command `$thing = 5` is an assignment operation. The number 5 is assigned or put into the container named *`$thing`*. In Perl, a variable can hold any kind of character string (numeric, alphabetic, or alphanumeric), but the names of variables must be prefixed with a dollar sign (for example, *`$thing`*).

vocabulary—Comprehensive collection of the words used in a general area of knowledge. Less focused than nomenclature.

Web services—Web services are resident server functionalities that accept client-initiated requests through an interface described by WSDL (Web services descriptive language).

`while (Perl)`—Evaluates a condition and loops through a block until the condition evaluates to false:

```
while ($line ne "")
    {
    $line = <TEXT>;
    print $line;
    }
```

`word barrier (Perl)`—\b is the place between a \w character and a \W character. It's a zero-width assertion, like ^, $, and the other anchors. It is not a character. A common misconception holds that \b is a synonym for \s+.

WSDL (Web Services Description Language)—An XML language used to describe a Web service's functionalities.

XML (eXtensible Markup Language)—An informatics technology that allows any data element (for example, a gene sequence, the weight of a patient, a biopsy diagnosis) to be bound to other data that describe the data element (metadata). It is surprising that this simple relationship between data and the data that describe data is the most powerful innovation in data organization since the invention of the book.

XML-RPC—An RPC (Remote Procedure call) rendered in XML syntax. XML-RPC is the forerunner of SOAP.

List of Perl Scripts

Author Biography

Jules Berman, PhD, MD, is a pathologist with an eclectic technical background that includes a bachelor's degree in mathematics from MIT. He was Program Director for Pathology Informatics in the Cancer Diagnosis Program at the U.S. National Cancer Institute for a period that spanned two millennia (1998-2005). He has first-authored more than 100 scientific papers and has recently authored another Jones & Bartlett book, "*Biomedical Informatics*" (2007). Dr. Berman is currently a freelance author and editor.

Index

U

V

W

X

Z